Collins

AQA GCSE
Maths

Higher Practice Book
Use and apply standard techniques

Rob Ellis

Contents (Higher tier only material appears **bold**)

How to use this book

Welcome to Collins *AQA GCSE Maths Higher Practice Book*. This book follows the structure of the Collins *AQA GCSE Maths 4th edition Higher Student Book*, so is ideal to use alongside it.

Colour-coded questions

Know what level of difficulty you are working at with questions ranging from more accessible (green), through intermediate (blue) to more challenging (pink).

Use of calculators

Questions when you could use a calculator are marked with a icon.

Hints and tips

These are provided where extra guidance can save you time or help you out.

Answers

Check your own work – the answers are provided online at www.collins.co.uk/gcsemaths4eanswers.

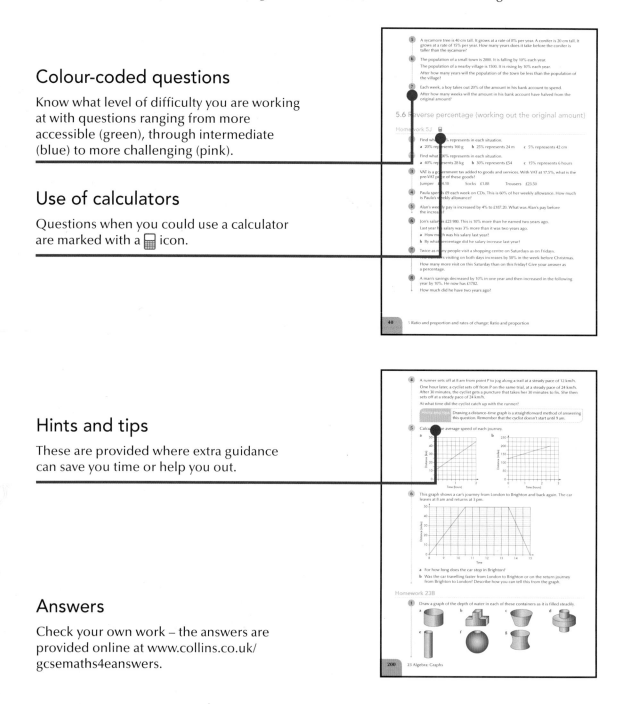

How to use this book

1 Number: Basic number

1.1 Solving real-life problems

Homework 1A

1 Andy needs enough tiles to cover an area of 12 m². 25 tiles cover 1 m². The tile store recommends buying 20 percent more tiles to allow for cutting.

Tiles are sold in boxes of 16. Andy buys 24 boxes.

Does he have enough tiles?

2 The organiser of a church fête needs 1000 balloons. Each packet contains 25 balloons and costs 85p. She has a budget of £30. Can she buy enough balloons?

3 A TV rental shop buys TVs for £110 each.

The shop needs to make a profit of at least 10% on each TV to cover its costs.

On average, each TV is rented for £3.50 per week for 40 weeks.

Does the shop cover its costs?

4 The annual membership fee for a fishing club is £42. The treasurer of the club has collected £1134 in fees.

How many people have paid their membership fees?

5 Mrs Woodhead saves £14 per week towards her bills. How much does she save each year?

6 Mark saves £15 each week for a dining set that costs £860.

Will he have saved enough money after one year? Show how you worked out your answer.

7 Sylvia has a part-time job and is paid £18 for every day she works. Last year she worked for 148 days. How much was she paid for the year?

8 Mutya has a part-time job, working three days each week.

She is paid £7 per hour and works for 4 hours each day.

Neil has a full-time job, working five days each week.

He is paid £8 per hour and works for 7 hours each day.

For how many weeks does Mutya have to work to earn at least as much as Neil earns in one week? Show your working.

9 A coach firm charges £504 for 36 people to go on a day trip to Calais. The cost of the coach is shared equally between the passengers.

Mary has £150 to pay for her trip and her shopping. When she gets to Calais, she wants to buy each of her four grandchildren a game that costs €50. The exchange rate is £1 = €1.25.

Does she have enough money?

10 A concert hall has 48 rows of seats with 32 seats in each row. What is the maximum capacity of the hall?

11 Allan is a market gardener. He plants 420 bulbs in rows of 18. How many complete rows will there be?

12 Sandeep wants a new carpet for her bedroom which is 6 m by 8 m. The carpet she has chosen costs £19 per square metre.

    a Estimate the cost of the carpet.   **b** Calculate the exact cost of the carpet.

13 Paul wants to carpet a room that measures 7 m by 3 m.

The carpet warehouse only stocks rolls that are 4 m wide and will only sell pieces that are the full width of the roll.

    a What is the smallest area of carpet that Paul can buy for his room?

    b Paul has £300 to spend on the carpet.

      What is the most he can spend per square metre?

14 240 students and teachers are going on a school trip.

The school has already booked four coaches that can each take 53 passengers.

They need to book one more coach.

What is the smallest number of seats needed on this coach?

15 On average, a toy shop sells 38 computer games each week.

The manager has a delivery of 150 games each month.

Will her stock of games increase or decrease? Show clearly how you decide.

1.2 Multiplication and division with decimals

Homework 1B

1 Round each number to the number of decimal places (dp) indicated.

    a 3.268 (1 dp)   **b** 0.0936 (2 dp)   **c** 64.815 93 (3 dp)   **d** 81.951 (2 dp)

    e 512.088 (1 dp)   **f** 954.672 (2 dp)   **g** 9.4364 (1 dp)   **h** 7.91373 (3 dp)

2 Work these out.

    a 0.5×0.5 **b** 12.6×0.6 **c** 7.2×0.7 **d** 1.4×1.2 **e** 2.6×1.5

3 For each part of this question:

    i estimate the answer by first rounding each number to the nearest whole number

    ii calculate the exact answer

    iii calculate the difference between your answers to parts **i** and **ii**.

    a 3.7×2.4 **b** 4.8×3.1 **c** 5.1×4.2 **d** 6.5×2.5

4   **a** Use any method to work out 15×16.

    b Use your answer to part a to work out each of these.

      i 1.5×1.6 **ii** 0.75×3.2 **iii** 4.5×1.6

5   **a** Work out 7.2×3.4.

    b Explain how you can use your answer to part **a** to write down the answer to 6.2×3.4.

6
  a Work out 2.3×7.5.
  b Use your answer to part **a** and the fact that $4.1 \times 7.5 = 30.75$, to work out 6.4×7.5.

7
  Evaluate each of these.
  a 3.12×14 **b** 5.24×15 **c** 1.36×22 **d** 7.53×25 **e** 27.1×32

8
  Work out the total cost of each purchase.
  a Twenty-four litres of petrol at £0.92 per litre
  b Eighteen pints of milk at £0.32 per pint
  c Fourteen magazines at £2.25 per copy

9
  A CD case is 0.8 cm thick.
  How many cases are in a pile of CDs that is 16 cm high?

10
  a Use any method to work out $64 \div 4$.
  b Use your answer to part a to work out each of these.
    i $6.4 \div 0.04$ **ii** $0.64 \div 4$ **iii** $0.064 \div 0.4$

11
  Here are three calculations.
  $43.68 \div 5.6$ $21.7 \div 6.2$ $19.74 \div 2.1$
  Which has the largest answer? Show how you know.

1.3 Approximation of calculations

Homework 1C

1
  Round each number to 1 significant figure.

a 51 203	**b** 56 189	**c** 33 261	**d** 89 998	**e** 94 999
f 53.71	**g** 87.24	**h** 31.06	**i** 97.835	**j** 184.23
k 0.5124	**l** 0.2765	**m** 0.006 12	**n** 0.049 21	**o** 0.000 888
p 9.7	**q** 85.1	**r** 91.86	**s** 196	**t** 987.65

2
  What are the least and the greatest numbers of people that live in these towns?
  a Hellaby   population 900 (to 1 significant figure)
  b Hook   population 650 (to 2 significant figure)
  c Hundleton   population 1050 (to 3 significant figures)

3
  Round each number to 2 significant figures.

a 6725	**b** 35 724	**c** 68 522	**d** 41 689	**e** 27 308
f 6973	**g** 2174	**h** 958	**i** 439	**j** 327.6

4
  Round each number to the number of significant figures (sf) indicated.

a 46 302 (1 sf)	**b** 6177 (2 sf)	**c** 89.67 (3 sf)	**d** 216.7 (2 sf)	**e** 7.78 (1 sf)
f 1.087 (2 sf)	**g** 729.9 (3 sf)	**h** 5821 (1 sf)	**i** 66.51 (2 sf)	**j** 5.986 (1 sf)
k 7.552 (1 sf)	**l** 9.7454 (3 sf)	**m** 25.76 (2 sf)	**n** 28.53 (1 sf)	**o** 869.89 (3 sf)
p 35.88 (1 sf)	**q** 0.084 71 (2 sf)	**r** 0.0099 (2 sf)	**s** 0.0809 (1 sf)	**t** 0.061 97 (3 sf)

5 A baker estimates that she has baked 100 loaves, to 1 significant figure.

She sells two loaves and now has 90 loaves, to 1 significant figure.

How many loaves did she start with? Work out all possible answers.

6 There are 500 cars in a car park, to 1 significant figure.

What is the least possible number of cars that could enter the car park so that there are 700 cars in the car park, to 1 significant figure?

7 Five minutes before closing time, a supermarket manager estimates that there are still 200 people shopping, to 1 significant figure.

No more shoppers can enter the supermarket.

On average, a checkout serves four customers in five minutes.

How many checkouts should be open so that all the customers can be served by closing time?

Homework 1D

1 Write down the answers, without using a calculator.

a 50×600
b 0.6×40
c 0.02×400
d $(30)^2$

e 0.5×250
f 0.6×0.7
g $30 \times 40 \times 50$
h $200 \times 0.7 \times 40$

2 Write down the answers, without using a calculator.

a $4000 \div 20$
b $8000 \div 200$
c $400 \div 0.5$
d $2000 \div 0.05$

e $1800 \div 0.12$
f $600 \div 0.3$
g $200 \times 30 \div 40$
h $300 \times 70 \div 0.4$

3 You are given that $18 \times 21 = 378$.

Use this information to write down the value of these calculations.

a 180×210
b $3780 \div 21$

4 Match each calculation with its answer and then write out the calculations in order, starting with the smallest answer.

6000×300 $\qquad$ 500×7000 $\qquad$ $10\,000 \times 900$ $\qquad$ $20 \times 80\,000$

$3\,500\,000$ $\qquad$ $1\,800\,000$ $\qquad$ $1\,600\,000$ $\qquad$ $9\,000\,000$

5 The Moon is approximately 400 000 km from Earth.

If a spaceship takes 8 days to travel to the Moon and return to Earth, how far does it travel each day?

Homework 1E

1 Work out approximate answers to each of these.

a 4324×6.71
b 6170×7.311
c 72.35×3.142

d 4709×3.81
e $63.1 \times 4.18 \times 8.32$
f $320 \times 6.95 \times 0.98$

g $454 \div 89.3$
h $26.8 \div 2.97$
i $4964 \div 7.23$

j $316 \div 3.87$
k $2489 \div 48.58$
l $63.94 \div 8.302$

2 By rounding each value in the calculation, find approximate answers to each of these.

a $\dfrac{561 \times 99}{101}$
b $\dfrac{491 - 210}{25}$
c $\dfrac{691 + 320}{989}$
d $\dfrac{59.1 \times 1.8}{2.56}$
e $\dfrac{9.1 \times 56}{18}$

1 Number: Basic number

3 Work out the approximate monthly pay of each person.

    a Joy: £47 200 per year   **b** Amy: £24 200 per year   **c** Tom: £19 135 per year

4 Work out the approximate annual pay of each person.

    a Trevor: £570 per week   **b** Brian: £2728 per month

5 A groundsman bought 350 kg of seed at a cost of £3.84 per kilogram. Find the approximate total cost of this seed.

6 By rounding each value in the calculation, find approximate answers to each of these.

    a $\dfrac{361 \times 89}{0.48}$ **b** $\dfrac{491 - 110}{0.18}$ **c** $\dfrac{211 + 420}{0.59}$ **d** $\dfrac{591 \times 18}{0.49}$ **e** $\dfrac{91 + 880}{0.67 - 0.58}$

7 A greengrocer sells a box of 250 apples for £47. If he sells them for 20p each, or more, he will make a profit.

    Does he make a profit? Use approximations to explain why.

8 Keith runs about 15 km every day. Approximately how far does he run in:

    a a week   **b** a month   **c** a year?

9 A litre of paint will cover an area of about 6.8 m². Approximately how many one-litre cans will I need to paint a fence with a total surface area of 43 m²?

10 A tour of London sets off at 10.13 am and costs £21. It returns at 12.08 pm.

    What is the approximate cost per hour of the tour?

11 Round each of the numbers in these statements to a suitable degree of accuracy.

    a Kris is 1.6248 m tall.

    b It took me 17 minutes 48.78 seconds to cook the dinner.

    c My rabbit weighs 2.867 kg.

    d The temperature at the bottom of the ocean is 1.239 °C.

    e There were 23 736 people at the baseball game yesterday.

12 How many jars, each holding 119 cm³ of water, can be filled from a three-litre flask?

13 If I walk at an average speed of 62 m per minute, how long will it take me to walk a distance of 4 km?

14 Helen earns £31 500 a year. She works 5 days a week for 45 weeks of the year.

    How much does she earn per day?

15 If 10 g of gold costs £2.17, how much will 1 kg of gold cost?

16 Rewrite this paragraph using sensible numbers.

    I left home at eleven and a half minutes past two and walked for 49 minutes. The temperature was 12.7623 °C. I could see an aeroplane overhead at 2937.1 feet. Altogether I walked 3.126 miles.

17 David travelled 350 miles in 5 hours 10 minutes.

    Trevor travelled half the distance in half the time.

    Approximately how fast was Trevor travelling?

1.4 Multiples, factors, prime numbers, powers and roots

Homework 1F

1 From the list of numbers below, write down the:

    a multiples of 4   **b** multiples of 5   **c** prime numbers   **d** factors of 2700.

      28   19   36   43   64   53   77

      66   56   60   15   29   61   45   51

2 During the peak travel time at a railway station, there are north-bound trains setting off every 8 minutes and south-bound trains setting off every 12 minutes. At 5 pm, one train sets off to the north and one train sets off to the south. How many more times will two trains set off at the same time before 6.30 pm?

3 Write down the negative square root of each number.

    a 36   **b** 81   **c** 100   **d** 900   **e** 361

    f 169   **g** 225   **h** 1 000 000   **i** 441   **j** 1225

4 Write down the cube root of each number.

    a 8   **b** 64   **c** 125   **d** 1000   **e** 27 000

    f −27   **g** −1   **h** −216   **i** −8000   **j** −343

5 Here are four numbers: 8, 20, 25, 64

Copy and complete this table by putting each number in the correct box.

	Square number	Factor of 40
Cube number		
Multiple of 5		

6 Use these four number cards to make a cube number.

     1   2   7   9

7 A number is a factor of 18 and a multiple of 18.

What is the number?

8 Write down the value of each expression.

    a $\sqrt{0.36}$ **b** $\sqrt{0.81}$ **c** $\sqrt{1.69}$ **d** $\sqrt{0.09}$ **e** $\sqrt{0.01}$

    f $\sqrt{1.44}$ **g** $\sqrt{2.25}$ **h** $\sqrt{1.96}$ **i** $\sqrt{4.41}$ **j** $\sqrt{12.25}$

1.5 Prime factors, LCM and HCF

Homework 1G

1 Draw prime factor trees for these numbers.

    a 144   **b** 75   **c** 98   **d** 420   **e** 560

2 Write these numbers as products of their prime factors using index notation.

    a 48   **b** 54   **c** 216   **d** 1000   **e** 675

3 a Express 36 as a product of its prime factors.

b Write your answer to part **a** using index notation.

c Use you answer to part **b** to write 18 and 72 as a product of their prime factors using index notation.

4 $119 = 7 \times 17$

$119^2 = 14\ 161$

$119^3 = 1\ 685\ 159$

a Write 14 161 as a product of prime factors using index notation.

b Write 1 685 159 as a product of prime factors using index notation.

c Write 119^{10} as a product of prime factors using index notation.

5 A mathematician wants to donate a total of £18 to three charities so that they each receive a whole number of pounds and the amount given to each charity is a factor of 18.

How much does he give to each charity?

6 The first three odd prime numbers are all factors of 105.

Explain why this means that seven people can share £105 equally so that each receives an exact number of pounds.

Homework 1H

1 Find the LCM of each pair of numbers.

a 5 and 7   b 3 and 8   c 6 and 9   d 10 and 12

e 10 and 15   f 12 and 16   g 16 and 24   h 15 and 35

2 Find the HCF of each pair of numbers.

a 21 and 49   b 27 and 45   c 15 and 25   d 25 and 45

e 48 and 60   f 72 and 108   g 54 and 126   h 99 and 132

3 Write each of these as a single power of x.

a $x^2 \times x^3$ b $x^4 \times x^5$ c $x^6 \times x$ d $x^5 \times x^5$ e $x^3 \times x^2 \times x^4$

4 Find the HCF of 55 555 and 67 750.

5 Find the LCM of 144 and 162.

6 Nuts are sold in packs of 12.

Bolts are sold in packs of 18.

What is the least number of each pack that needs to be bought in order to have the same numbers of nuts and bolts?

7 The HCF of two numbers is 5.

The LCM of the same two numbers is 150.

What are the numbers?

1.6 Negative numbers

Homework 1I

1
  a Work out $17 \times (-4)$.

  b The average temperature drops by 4 °C each day for 17 days. By how much has the temperature dropped altogether?

  c The temperature drops by 6 °C for each of the next four days. Write down the calculation to work out the total change in temperature over these four days.

2 Write down the answer to each calculation.

a -2×4	**b** -3×6	**c** -5×7	**d** $-3 \times (-4)$	**e** $-8 \times (-2)$
f $-14 \div (-2)$	**g** $-16 \div (-4)$	**h** $25 \div (-5)$	**i** $-16 \div (-8)$	**j** $-8 \div (-4)$
k $3 \times (-7)$	**l** $6 \times (-3)$	**m** $7 \times (-4)$	**n** $-3 \times (-9)$	**o** $-7 \times (-2)$
p $28 \div (-4)$	**q** $12 \div (-3)$	**r** $-40 \div 8$	**s** $-15 \div (-3)$	**t** $50 \div (-2)$
u $-3 \times (-8)$	**v** $42 \div (-6)$	**w** $7 \times (-9)$	**x** $-24 \div (-4)$	**y** -7×8

3 Write down the answer to each calculation.

a $-2 + 4$	**b** $-3 + 6$	**c** $-5 + 7$	**d** $-3 + (-4)$	**e** $-8 + (-2)$
f $-14 - (-2)$	**g** $-16 - (-4)$	**h** $25 - (-5)$	**i** $-16 - (-8)$	**j** $-8 - (-4)$
k $3 + (-7)$	**l** $6 + (-3)$	**m** $7 + (-4)$	**n** $-3 + (-9)$	**o** $-7 + (-2)$
p $28 - (-4)$	**q** $12 - (-3)$	**r** $-40 - 8$	**s** $-15 - (-3)$	**t** $50 - (-2)$
u $-3 + (-8)$	**v** $42 - (-6)$	**w** $7 + (-8)$	**x** $-24 - (-4)$	**y** $-7 + 8$

4 By what number do you multiply -5 by to get each of these numbers?

  a 25   **b** -30 **c** 50   **d** -100 **e** 75

5 Put these calculations in order from lowest result to highest.

  $-18 \div 12$ $-0.5 \times (-4)$ $-21 \div (-14)$ $0.3 \times (-2)$

Homework 1J

1 Work out each of these. Remember to work out the brackets first.

a $-3 \times (-2 + 6)$	**b** $8 \div (-3 + 2)$	**c** $(6 - 8) \times (-3)$	**d** $-4 \times (-6 - 3)$
e $-5 \times (-6 \div 2)$	**f** $(-5 + 3) \times (-3)$	**g** $(6 - 9) \times (-4)$	**h** $(2 - 5) \times (5 - 2)$

2 Work these out.

a $-5 \times (-4) + 3$	**b** $-8 \div 8 - 3$	**c** $16 \div (-4) + 3$	**d** $2 \times (-5) + 6$
e $-3 \times 4 - 5$	**f** $-1 + 42 - 5$	**g** $5 - 32 + 2$	**h** $-1 + 2 \times (-3)$

3 Copy each of these and put in brackets where necessary to make each one true.

  a $4 \times (-3) + 2 = -4$ **b** $-6 \div (-3) + 2 = 4$ **c** $-6 \div (-3) + 2 = 6$

4 Work out the value of each expression when $a = -3$, $b = 5$ and $c = -4$.

  a $(a + b)^2$ **b** $-(a + c)^2$ **c** $(a + b)c$ **d** $a^2 + b^2 + c^2$

5 Work out the value of each expression.

  a $(7^2 + 2^2) \times 3$ **b** $18 \div (2 - 5)^2$ **c** $3 \times (6^2 - (1 - 8)^2)$ **d** $((7 + 1)^2 - (2 - 3)^2) \div 7$

6 Use each of the numbers 4, 6 and 8 and each of the symbols $-$, $\times$ and $\div$ to make a calculation with an answer -3.

7 Use any four different numbers to make a calculation with an answer of -9.

8 Use the numbers 1, 2, 3, 4 and 5 in order, from smallest to largest, together with one of each of the symbols $+$, $-$, $\times$, $\div$ and two pairs of brackets to make a calculation with an answer of -2.75.

  For example, to make a calculation with an answer of -4: $(1 + 2) - (3 \times 4) + 5 = -4$.

2 Number: Fractions, ratio and proportion

2.1 One quantity as a fraction of another

Homework 2A

1. Write the first quantity as a fraction of the second.

   a 5p, 20p   b 4 kg, 12 kg   c 6 hours, 12 hours

   d 14 days, 30 days   e 6 days, 2 weeks   f 20 minutes, 2 hours

2. David scored 18 goals out of his team's season total of 48 goals. What fraction of all the goals did David score?

3. Angeliki works for 32 weeks a year. What fraction of the year does she work for?

4. Mark earns £120 and saves £40 of it.

   Bev earns £150 and saves £60 of it.

   Who is saving the greater proportion of their earnings?

5. In a test, Kevin scores 7 out of 10 and Sally scores 15 out of 20.

   Which is the better mark? Explain your answer.

6. In a dance troupe of 20 dancers, 12 are women. The rest are men. Half of the men are single.

   What fraction of the dancers are single men? Give your answer in its simplest form.

7. A bus driver says that at least two out of every three passengers on her bus have a travel pass. The conductor says it is less than three out of four.

   If both statements are true and the bus is carrying 50 passengers, how many have a travel pass? Write down all possible answers.

2.2 Adding, subtracting and calculating with fractions

Homework 2B

1. Work these out.

   a $\dfrac{1}{2}+\dfrac{1}{5}$ b $\dfrac{1}{2}+\dfrac{1}{3}$ c $\dfrac{1}{3}+\dfrac{1}{10}$ d $\dfrac{3}{8}+\dfrac{1}{3}$

   e $\dfrac{3}{4}+\dfrac{1}{5}$ f $\dfrac{1}{3}+\dfrac{2}{5}$ g $\dfrac{3}{5}+\dfrac{3}{8}$ h $\dfrac{1}{2}+\dfrac{2}{5}$

2. Evaluate the following.

   a $\dfrac{1}{2}+\dfrac{1}{4}$ b $\dfrac{1}{3}+\dfrac{1}{6}$ c $\dfrac{3}{5}+\dfrac{1}{10}$ d $\dfrac{5}{8}+\dfrac{1}{4}$

3 Work these out.

a $\dfrac{7}{8} - \dfrac{3}{4}$ b $\dfrac{4}{5} - \dfrac{1}{2}$ c $\dfrac{2}{3} - \dfrac{1}{5}$ d $\dfrac{3}{4} - \dfrac{2}{5}$

4 Evaluate the following.

a $\dfrac{5}{8} + \dfrac{3}{4}$ b $\dfrac{1}{2} + \dfrac{3}{5}$ c $\dfrac{5}{6} + \dfrac{1}{4}$ d $\dfrac{2}{3} + \dfrac{3}{4}$

5 a At a football club, half of the players are English, a quarter are Scottish and one-sixth are Italian. The rest are Irish. What fraction of players at the club are Irish?

b One of the following is the number of players at the club.

30   32   34   36

How many players are at the club?

6 On a firm's coach trip, half the people were employees and two-fifths were partners of the employees. The rest were children. What fraction were children?

7 Five-eighths of a crowd of 35 000 people were male. How many females were in the crowd?

8 What is four-fifths of 65 added to five-sixths of 54?

9 Which three of these four fractions add up to 1?

$\dfrac{1}{6}$ $\dfrac{5}{12}$ $\dfrac{1}{4}$ $\dfrac{1}{3}$

10 Pipes are sold in $\frac{1}{2}$-m and $\frac{3}{4}$-m lengths.

What is the least number of pipes that can be used to make a pipe exactly 2 m long? (Assume that you cannot cut pipes to size.)

Show your working.

2.3 Multiplying and dividing fractions

Homework 2C

1 Work these out. Give each answer in its simplest form.

a $\dfrac{1}{2} \times \dfrac{2}{3}$ b $\dfrac{3}{4} \times \dfrac{2}{5}$ c $\dfrac{3}{5} \times \dfrac{1}{2}$ d $\dfrac{3}{7} \times \dfrac{2}{3}$ e $\dfrac{2}{3} \times \dfrac{5}{6}$

f $\dfrac{1}{3} \times \dfrac{3}{5}$ g $\dfrac{2}{3} \times \dfrac{7}{10}$ h $\dfrac{3}{8} \times \dfrac{2}{5}$ i $\dfrac{4}{9} \times \dfrac{3}{8}$ j $\dfrac{4}{5} \times \dfrac{7}{16}$

2 Work these out. Give your answer as a mixed number where possible.

a $\dfrac{1}{5} \div \dfrac{1}{3}$ b $\dfrac{3}{5} \div \dfrac{3}{8}$ c $\dfrac{4}{5} \div \dfrac{2}{3}$ d $\dfrac{4}{7} \div \dfrac{8}{9}$

e $4 \div 1\dfrac{1}{2}$ f $5 \div 3\dfrac{2}{3}$ g $8 \div 1\dfrac{3}{4}$ h $6 \div 1\dfrac{1}{4}$

i $5\dfrac{1}{2} \div 1\dfrac{1}{3}$ j $7\dfrac{1}{2} \div 2\dfrac{2}{3}$ k $1\dfrac{1}{2} \div 1\dfrac{1}{5}$ l $3\dfrac{1}{5} \div 3\dfrac{3}{4}$

3 Kris walked three-quarters of the way along Carterknowle Road which is 3 km long. How far did Kris walk?

4 Jean ate one-fifth of a cake, Les ate half of what was left. Nick ate the rest. What fraction of the cake did Nick eat?

5 A formula for working out the distance travelled is:

distance travelled = speed × time taken.

A snail is moving at one-tenth of a metre per minute. It travels for half a minute.

How far has it travelled?

6 George is given £80. Each week he spends half of the amount left.

What fraction of the £80 will he have left after 4 weeks?

7 You are given that 1 litre = 100 cl.

A bottle holds 1.5 litres of water.

Tupac drinks half of the water.

Belinda drinks 25 cl of the water.

What fraction of the contents is left?

8 Work these out. Give your answer as a mixed number where possible.

a $1\frac{1}{3} \times 2\frac{1}{4}$ **b** $1\frac{3}{4} \times 1\frac{1}{3}$ **c** $2\frac{1}{2} \times \frac{4}{5}$ **d** $1\frac{2}{3} \times 1\frac{3}{10}$

e $3\frac{1}{4} \times 1\frac{3}{5}$ **f** $2\frac{2}{3} \times 1\frac{3}{4}$ **g** $3\frac{1}{2} \times 1\frac{1}{6}$ **h** $7\frac{1}{2} \times 1\frac{3}{5}$

9 Write down the reciprocal of each number.

a 5   **b** $\frac{1}{7}$ **c** $\frac{2}{5}$ **d** $\frac{9}{5}$

10 Write down the negative reciprocal of each number.

a 3   **b** $-\frac{1}{4}$ **c** $\frac{7}{11}$ **d** $-\frac{4}{5}$

11 Which is smaller: $\frac{3}{4}$ of $5\frac{1}{3}$ or $\frac{2}{3}$ of $4\frac{2}{5}$?

12 I estimate that I need 60 litres of lemonade for a party.

I buy 24 bottles, each containing $2\frac{3}{4}$ litres.

Have I bought enough lemonade for the party?

13 Pizzas are often cut into eight equal pieces.

If a pizza is cut into six equal pieces how much more pizza is in each piece?

Give your answer as a fraction of the whole pizza.

14 A company employs 200 people.

The manager says that exactly two-thirds of the employees are women and three-quarters of the employees are full-time.

One of the statements is true and one is not accurate.

Explain which statement is which.

15 A fish farmer is trying to work out how many fish are in a pond.

He captures 100 fish, marks them and puts them back in the pond.

Later he captures 100 fish and finds that 25 are marked.

Approximately how many fish are in the pond?

16 A pet shop has 36 kg of hamster food. The owner wants to pack this into bags, each containing three-quarters of a kilogram. How many bags can he make in this way?

17 Sergio wants to put up a fence along down one side of his garden which is 20 m long. The fence comes in sections; $1\frac{1}{3}$ m long. How many sections will Sergio need to put the fence all the way down his garden?

18 An African bullfrog can jump a distance of $1\frac{1}{4}$ m in one hop. How many hops would it take an African bullfrog to hop a distance of 100 m?

19 Three-fifths of all 14-year-olds in a school visit the dentist each year.

One-third of those who do not visit the dentist have a problem with their teeth.

What fraction of all the 14-year-olds do not visit the dentist and have a problem with their teeth?

20 How many half-litre tins of paint can be poured into a 2.2 litre paint tray without spilling?

21 I work 8 hours each day.

Short tasks last $\frac{1}{4}$ hour.

Long tasks last $\frac{3}{4}$ hour.

I have to complete at least three long tasks.

How many short tasks can I also complete in one working day?

22 Work these out. Give your answer as a mixed number where possible.

a $\frac{4}{5} \times \frac{1}{2} \times \frac{3}{8}$ b $\frac{3}{4} \times \frac{7}{10} \times \frac{5}{6}$ c $\frac{2}{3} \times \frac{5}{6} \times \frac{9}{10}$

d $1\frac{1}{4} \times \frac{2}{3} \div \frac{5}{6}$ e $\frac{5}{8} \times 1\frac{1}{4} \div 1\frac{1}{10}$ f $2\frac{1}{2} \times 1\frac{1}{3} \div 3\frac{1}{3}$

2.4 Fractions on a calculator

Homework 2D

1 Use your calculator to work these out. Give your answers as fractions.

a $\frac{3}{4} + \frac{2}{3}$ b $\frac{4}{5} + \frac{7}{10}$ c $\frac{3}{8} + \frac{1}{4} + \frac{2}{5}$

d $\frac{9}{10} - \frac{5}{12}$ e $\frac{7}{12} + \frac{5}{8} - \frac{1}{6}$

2 Use your calculator to work these out. Give your answers as mixed numbers.

a $4\frac{3}{5} + 2\frac{3}{4}$ b $3\frac{1}{6} + 4\frac{4}{5}$ c $1\frac{5}{8} + 3\frac{5}{16} + 3\frac{1}{24}$

d $6\frac{7}{12} - 4\frac{1}{8}$ e $6\frac{7}{16} + 3\frac{3}{7} - 7\frac{7}{20}$

3 a Use your calculator to work out $\frac{31}{43} - \frac{29}{125}$.

4 A shape is rotated 80° clockwise and then a further 40° clockwise.

What fraction of a turn will return it to its original position?

Give both possible answers.

5 Use your calculator to work these out. Give your answers as fractions.

a $\frac{3}{7} \times \frac{3}{4}$ b $\frac{4}{9} \times \frac{10}{11} \times \frac{3}{8}$ c $\frac{9}{10} \div \frac{3}{5}$

d $\frac{3}{4} \div \frac{5}{24}$ e $\frac{5}{12} \times \frac{6}{11} \div \frac{1}{30}$ f $\frac{3}{5} \times \frac{2}{13} \div \frac{3}{11}$

2 Number: Fractions, ratio and proportion

6 Use your calculator to work these out. Give your answers as mixed numbers.

a $3\frac{3}{4} \times 2\frac{3}{10}$ **b** $9\frac{3}{5} \times 11\frac{6}{11}$ **c** $4\frac{3}{8} \times 3\frac{7}{16} \times 6\frac{2}{5}$

d $5\frac{7}{20} \div 2\frac{5}{12}$ **e** $7\frac{3}{5} \div 3\frac{7}{18}$

7 The formula for the area of a rectangle is area = length × width.

Use this formula to work out the area of a rectangle of length $7\frac{2}{5}$ metres and width $4\frac{1}{4}$ metres.

8 A length of wood is $4\frac{1}{2}$ m long. Angharad wants to cut pieces of wood that are each $\frac{1}{8}$ m long.

How pieces of wood can she cut from this length?

2.5 Increasing and decreasing quantities by a percentage

Homework 2D

1 Work out the multiplier you would use to increase a quantity by each percentage.

a 7%   **b** 2%   **c** 30%   **d** 6%   **e** 15%

2 Work out the multiplier you would use to decrease a quantity by each percentage.

a 9%   **b** 14%   **c** 35%   **d** 12%   **e** 22%?

3 Increase each amount by the given percentage.

a £80 by 5%   **b** 14 kg by 6%   **c** £42 by 3%

4 Increase each amount by the given percentage.

a 340 g by 10%   **b** 64 m by 5%   **c** £41 by 20%

5 Keith was on a salary of £34 200. He was given a pay rise of 4%. What is his new salary?

6 In 2004 the population of Dripfield was 14 200. By 2014, it had increased by 8%. What was the population of Dripfield in 2014?

7 In 2003 the number of bikes on the roads of Doncaster was about 840. Since then it has increased by 8%. Approximately how many bikes are now on the roads of Doncaster?

8 A dining table costs £300 before the VAT is added.

If the rate of VAT increases from 15% to 20%, by how much will the cost of the dining table increase?

9 A restaurant meal is advertised at £20. A service charge is added to all bills.

The bill for two people is shown.

Show that the service charge is 15%.

Romano's Bistro	
2 × Prawn Cocktail	4.60
1 × Chicken Risotto	6.60
1 × Sea Bass	8.50
1 × Fries	4.50
1 × Tiramisu	6.60
1 × Choc fudge cake	4.20
2 × Coffee	5.00
Service charge	6.00
Total	46.00

10 A shopkeeper decides to increase prices by 5% or £1, whichever is greater.

What is the price range of the items that will increase by £1?

11 Decrease each amount by the given percentage. (Use any method you like.)

a £20 by 10%   b £150 by 20%   c 90 kg by 30%   d 500 m by 12%

e £260 by 5%   f 80 cm by 25%   g 400 g by 42%   h £425 by 23%

i 48 kg by 75%   j £63 by 37%

12 Mrs Denghali buys a new car from a garage for £8400. The garage owner tells her that her car will lose 24% of its value after one year. What will be the value of the car after one year?

13 In 2010 the population of a village was 2400. In 2014 the population had decreased by 12%. What was the population of the village in 2014?

14 A Travel Agent is offering a 15% discount on holidays. How much will the advertised holiday now cost?

NEW YORK
FOR A WEEK   **£540**

15 Matt was given £160 at Christmas. In a New Year's Sale, all prices are reduced by 20%. Can he afford to buy a shirt that normally costs £30, a suit that normally costs £130 and a pair of shoes that normally cost £42?

NEW YEAR'S SALE:
All prices reduced by 20%

16 On the first day of a new term, a school expects to have an attendance rate of 99%. If the school population is 700 students, how many students will the school expect to be absent on the first day of the new term?

17 Putting cavity wall insulation into your home reduces fuel use by 20%. A family using an average of 850 units of electricity a year put cavity wall insulation into their home. How much electricity would they expect to use now?

18 A shop increases all its prices by 10%.

One month later it advertises 10% off all marked prices.

Are the goods cheaper, the same price or more expensive than before the increase?

Show how you work out your answer.

19 Show that a 10% increase followed by a 10% decrease is equivalent to a 1% decrease overall.

2.6 Expressing one quantity as a percentage of another

Homework 2F

1 Write the first quantity as a percentage of the second. Give suitably rounded figures where necessary.

a £8, £40   b 20 kg, 80 kg   c 5 m, 50 m   d £15, £20

e 400 g, 500 g   f 23 cm, 50 cm   g £12, £36   h 18 minutes, 1 hour

i £27, £40   j 5 days, 3 weeks

2 What percentage of these shapes is shaded?

a

b

3 In a class of 30 students, 18 are girls.

    a What percentage of the class are girls?   **b** What percentage of the class are boys?

4 The area of a farm is 820 hectares. The farmer uses 240 hectares for pasture.

What percentage of the farm land is used for pasture? Give your answer to one decimal place.

5 Find the percentage profit on each item. Give your answers to one decimal place.

	Item	Retail price or selling price (£)	Wholesale price paid by the shop (£)
a	Micro hi-fi system	250	150
b	Mp3 player	90	60
c	CD player	44.99	30
d	Cordless headphones	29.99	18

6 These are the results from two tests taken by Paul and Val. Both tests are out of the same mark.

Whose result has the greater percentage increase from test A to test B?

Show your working.

	Test A	Test B
Paul	30	40
Val	28	39

7 A small train is carrying 48 passengers.

At a station, more passengers get on so that all seats are filled and no one is standing.

At the next station, 70% of the passengers leave the train and 30 new passengers get on.

There are now 48 passengers on the train again.

How many seats are on the train?

8 In a secondary school, 30% of students have a younger brother or sister at primary school.

20% of students have two younger brothers or sisters at primary school

Altogether there are 700 brothers and sisters at the primary school.

How many students are at the secondary school?

9 James came home from school with his end-of-year test results. Change each of James' results to a percentage.

    Maths   63 out of 75   English   56 out of 80

    Science   75 out of 120   French   27 out of 60

10 In 1993, it rained in London on 80 days of the year. What percentage of the days of the year were wet? Give your answer to two significant figures.

3 Statistics: Statistical diagrams and averages

3.1 Statistical representation

Homework 3A

1 The table shows the times taken by 60 people to travel to work.

Time in minutes	10 or less	Between 10 and 30	30 or more
Frequency	8	19	33

Draw a pie chart to illustrate the data.

2 The table shows the numbers of GCSE passes that 180 students obtained.

GCSE passes	9 or more	7 or 8	5 or 6	4 or less
Frequency	20	100	50	10

Draw a suitable chart to illustrate the data.

3 Marion is writing an magazine article on health. She asked a sample of people the question: "How often do you consider your health when planning your diet?" The pie chart shows the results of her survey.

a What percentage of the sample responded 'often'.

b What response was given by about a third of the sample?

c Can you tell how many people there were in the sample? Give a reason for your answer.

d What other questions could Marion ask?

4 Tom is doing a statistics project on the use of computers. He asks 36 of his school friends about their main use of computers and records the results in the table shown below.

Main use	e-mail	Internet	Word processing	Games
Frequency	5	13	3	15

a Draw a pie chart to illustrate his data.

b What conclusions can you draw from his data?

c Give reasons why Tom's data is not really suitable for his project.

5 In a survey, a TV researcher asks 120 people at a leisure centre to name their favourite type of television programme. The results are shown in the table.

Type of programme	Comedy	Drama	Films	Soaps	Sport
Frequency	18	11	21	26	44

a Draw a pie chart to illustrate the data.

b Do you think the sample chosen by the researcher is representative of the population? Give a reason for your answer.

Homework 3B

1 The table shows the estimated number of visitors to the cinema in Sheffield.

Year	1975	1980	1985	1990	1995	2000	2005	2010
Number of visitors (thousands)	280	110	180	330	510	620	750	810

a Draw a line graph for the data.

b Use your graph to estimate the number of visitors to the cinema in 2003.

c In which five-year period did the number of visitors increase the most?

d Explain the trend in the number of visitors. What reasons can you give to explain this trend?

2 Callum opened a new tea shop and was interested in how trade was picking up over the first few weeks. The table shows the number of teas sold in these weeks.

Week	1	2	3	4	5
Teas sold	67	82	100	114	124

a Draw a line graph for this data.

b From your graph, estimate the number of teas Callum can hope to see in week 6.

c Give a possible reason for the way in which the number of teas sold increases?

3 A kitten is weighed at the end of each week, for five weeks after it is born.

Week	1	2	3	4	5
Mass (g)	420	480	530	560	580

a Estimate how much the kitten would weigh after 8 weeks.

b Why might this not be a sensible estimate?

4 When plotting a graph to show the winter midday temperatures in Mexico, Pete decided to start his graph at the temperature 10°C.

Explain why he might have made this decision.

3.2 Statistical measures

Homework 3C

1 **a** Find the mode, the median and the mean for each set of data.

    i   6, 4, 5, 6, 2, 3, 2, 4, 5, 6, 1   **ii**   14, 15, 15, 16, 15, 15, 14, 16, 15, 16, 15

    iii 31, 34, 33, 32, 46, 29, 30, 32, 31, 32, 33

    b For each set of data, decide which average you would use. Give a reason for each answer.

2 A supermarket sells oranges in bags of ten.

    The masses of each orange in a selected bag were as follows:

    134 g, 135 g, 142 g, 153 g, 156 g, 132 g, 135 g, 140 g, 148 g, 155 g.

    a Find the mode, the median and the mean for the mass of the oranges.

    b The supermarket wanted to state the average mass on each bag they sold. Which of the three averages would you advise the supermarket to use? Explain why.

3 The mass, in kilograms, of players in a school football team are listed below.

    68, 72, 74, 68, 71, 78, 53, 67, 72, 77, 70

    a Find the median mass of the team.   **b** Find the mean mass of the team.

    c Which average is the better one to use? Explain why.

4 Jez is a member of a local quiz team. The number of points he has scored in the last eight weeks are listed below.

    62, 58, 24, 47, 64, 52, 60, 65

    a Find the median for the number of points he scored over the eight weeks.

    b Find the mean for the number of points he scored over the eight weeks.

    c The team captain wanted to know the average for each member of the team. Which average would Jez use? Give a reason for your answer.

5 Three dancers were hoping to be chosen to represent their school in a competition.

    The table below shows their scores in some recent competitions.

Kathy	8, 5, 6, 5, 7, 4, 5
Connie	8, 2, 7, 9, 2
Evie	8, 1, 8, 2, 3

    The teacher said they would be selected by their best average score.

    By which average would each dancer choose to be selected?

6 **a** Find three numbers that have a range of 3 and a mean of 3.

    b Find three numbers that have a range of 3, a median of 3 and mean of 3.

7 A class of students took a test.

    When talking about the results, the teacher said the average mark was 32. One of the students said it was 28.

    Explain how they could both be correct.

1 Work out: **i** the mode **ii** the median **iii** the mean from each frequency table.

a The results of a survey of the collar sizes of all male staff

Collar size	12	13	14	15	16	17	18
Number of staff	1	3	12	21	22	8	1

b The results of a survey of the number of TVs in students' homes

Number of TVs	1	2	3	4	5	6	7
Frequency	12	17	30	71	96	74	25

2 A survey of the number of pets in each family of a school gave these results.

Number of pets	0	1	2	3	4	5
Frequency	28	114	108	16	15	8

a Assuming each child at the school was counted in the data, how many children are at the school?

b Calculate the median number of pets in a family.

c How many families have less than the median number of pets?

d Calculate the mean number of pets in a family. Give your answer to 1 dp.

3 The table shows the number of days each week that Bethan travelled to Manchester on business.

Days	0	1	2	3	4	5
Number of weeks	17	2	4	13	15	1

Explain how you would find the median number of days per week that Bethan travelled to Manchester.

4 A survey of the number of television sets in each family home in one school year gave these results.

Number of TVs	0	1	2	3	4	5
Frequency	1	5	36	86	72	56

a How many students are in that school year?

b Calculate the mean number of TVs in a home.

c How many homes have this mean number of TVs (if you round the mean to the nearest whole number)?

d What percentage of homes could consider themselves average from this survey?

5 The number of eggs laid by 20 hens one day is shown in frequency table, but a coffee stain has obscured two columns of data.

Eggs	0	1	2		5
Frequency	2	3	4		1

The mean number of eggs laid was 2.5.

What could the four missing numbers be?

1 For each table of values, find: **i** the modal group **ii** an estimate for the mean.

a

Score	0–20	21–40	41–60	61–80	81–100
Frequency	9	13	21	34	17

b

Cost (£)	0.00–10.00	10.01–20.00	20.01–30.00	30.01–40.00	40.01–50.00
Frequency	9	17	27	21	14

2 A hospital has to report the average waiting time for patients in the Accident and Emergency department. During one shift, a survey was made to see how long these patients had to wait before seeing a doctor. The table summarises the results.

Time (minutes)	0–10	11–20	21–30	31–40	41–50	51–60	61–70
Frequency	1	12	24	15	13	9	5

a How many patients were seen by a doctor during this shift?

b Estimate the mean waiting time.

c Which average would the hospital use to represent the waiting time?

d What percentage of patients did the doctors see within the hour?

3 A shop recorded the sizes of men's shoes sold in one month. The table summarises the results.

Shoe size	3–4	5–6	7–8	9–10	11–12
Frequency	2	4	21	55	32

a How many pairs of men's shoes were sold during this month?

b Estimate the mean size of men's shoe sold.

c Which of the averages is of most useful for the shop manager?

d What percentage of men's shoes sold were smaller than size 7?

4 The table below shows the total mass of fish caught by the anglers in a fishing competition.

Mass, m (kg)	$0 \leqslant m < 5$	$5 \leqslant m < 10$	$10 \leqslant m < 15$	$15 \leqslant m < 20$	$20 \leqslant m < 25$
Frequency	4	15	10	8	3

Chloe noticed that two numbers were in the wrong part of the table and that this made a difference of 0.625 to the arithmetic mean.

Which two numbers were the wrong way round?

5 The profit made each week by a tea shop is shown in the table below.

Profit	£0–£200	£201–£400	£401–£600	£601–£800
Frequency	15	26	8	3

Explain how you would estimate the mean profit made each week.

3.3 Scatter diagrams

Homework 3F 🖩

1   The table shows the heights and masses of 12 students in a class.

Student	Mass (kg)	Height (cm)
Arianna	41	143
Bea	48	145
Caroline	47.5	147
Dhiaan	52	148
Emma	49.5	149
Fiona	55	149
Gill	55	153
Hanna	55.5	155
Imogen	61	157
Jasmine	65.5	160
Keira	60	163
Laura	68	165

   a Plot the data on a scatter diagram.

   b Draw the line of best fit.

   c Chloe was absent from the class, but is 152 cm tall. Use the line of best fit to estimate her mass.

   d A new girl, with a mass of 45 kg, joined the class. What height would you expect her to be?

2   The table shows the marks for 10 students in their mathematics and music examinations.

Student	Alex	Ben	Chris	Dom	Ellie	Ffion	Giordan	Hannah	Isabel	Jemma
Maths	52	42	65	60	77	83	78	87	29	53
Music	50	52	60	59	61	74	64	68	26	45

   a Plot the data on a scatter diagram. Use the horizontal axis for the mathematics scores and mark it from 20 to 100. Use the vertical axis for the music scores and mark it from 20 to 100.

   b Draw a line of best fit.

   c One of the students was ill when they took the music examination. Which student was it most likely to have been?

   d Another student, Kris, was absent for the music examination but scored 45 in mathematics. What mark would you expect him to have scored in music?

   e Lexie was absent for the mathematics examination but scored 78 in music. What mark would you expect her to have got in mathematics?

3   The table shows the time taken and distance travelled by a delivery van for 10 deliveries in one day.

Distance (km)	8	41	26	33	24	36	20	29	44	27
Time (minutes)	21	119	77	91	63	105	56	77	112	70

   a Draw a scatter diagram with time on the horizontal axis.

   b Draw a line of best fit on your diagram.

   c A delivery takes 45 minutes. How many kilometres would you expect the journey to have been?

   d How long would you expect a journey of 30 kilometres to take?

4   Harry records the time taken, in hours, and the distance travelled, in miles, for several different journeys.

Time (hours)	1	1.6	2.2	2.6	3.2	3.5	4	4.8	5.2
Distance (miles)	42	62	86	104	130	105	165	190	210

   Estimate the distance travelled for a journey lasting 200 minutes.

5   Describe what you would expect the scatter graph to look like if someone said that it showed positive correlation.

4 Algebra: Number and sequences

4.1 Patterns in number

Homework 4A

For questions **1** to **5**, copy and complete the number sentences given, look for the pattern and then write the next three lines. Write down anything you notice about the patterns.

1 $2 \times 11 =$ $\qquad$ $22 \times 11 =$ $\qquad$ $222 \times 11 =$

2 $99 \times 11 =$ $\qquad$ $999 \times 11 =$ $\qquad$ $9999 \times 11 =$

3 $7 \times 9 =$ $\qquad$ $7 \times 99 =$ $\qquad$ $7 \times 999 =$

4 $11 = 11$ $\qquad$ $11 \times 11 = 121$ $\qquad$ $11 \times 11 \times 11 =$

5 $9 \times 2 =$
$9 \times 3 =$
$9 \times 4 =$
$9 \times 5 =$

6 This is part of Pascal's Triangle.

    1
  1 2 1
1 3 3 1

  a Complete the next five lines of the triangle.

  b Describe how each line is formed.

  c Highlight any patterns you can see in the triangle.

4.2 Number sequences

Homework 4B

1 Write down the next three terms in each sequence and describe the pattern.

  a 4, 6, 8, 10, …   **b** 3, 6, 9, 12, …   **c** 2, 4, 8, 16, …

  d 5, 12, 19, 26, …   **e** 3, 30, 300, 3000, …   **f** 1, 4, 9, 16, …

2 Write down the next two terms in each sequence and describe the pattern.

  a 1, 1, 2, 3, 5, 8, 13, 21, …   **b** 2, 3, 5, 8, 12, 17, …

3 For each number sequence, find the rule and write down the next three terms.

    a 7, 14, 28, 56, …   **b** 3, 10, 17, 24, 31, …   **c** 1, 3, 7, 15, 31, …

    d 40, 39, 37, 34, …   **e** 3, 6, 11, 18, 27, …   **f** 4, 5, 7, 10, 14, 19, …

    g 4, 6, 7, 9, 10, 12, …   **h** 5, 8, 11, 14, 17, …   **i** 5, 7, 10, 14, 19, 25, …

    j 10, 9, 7, 4, …   **k** 200, 40, 8, 1.6, …   **l** 3, 1.5, 0.75, 0.375, …

4 A fraction sequence is formed by the rule $\frac{n+1}{2n+1}$.

Show that in the first eight terms, only one of the fractions is a terminating decimal.

5 A physiotherapist uses the expressions below for charging for a series of n sessions, when they are paid for in advance.

For $n < 5$, cost is £$(35n + 20)$

For $6 < n < 10$, cost is £$(35n + 10)$

For $n > 11$, cost is £$35n$

    a How much will the physiotherapist charge for 8 sessions booked in advance?

    b How much will the physiotherapist charge for 14 sessions booked in advance?

    c One client paid £220 in advance for a series of sessions.

       How many sessions did she book?

    d A runner with a leg injury books four sessions. After the sessions, he starts to run in races again. The leg injury returns and he has to book three more sessions.

       How much more does he pay for his treatment than he would do if he booked all the sessions at the same time?

4.3 Finding the nth term of a linear sequence

Homework 4C

1 Find the nth term in each linear sequence.

    a 5, 7, 9, 11, 13 …   **b** 3, 11, 19, 27, 35, …   **c** 6, 11, 16, 21, 26, …

    d 3, 9, 15, 21, 27, …   **e** 4, 7, 10, 13, 16, …   **f** 3, 10, 17, 24, 31, …

2 Find the 50th term in each linear sequence.

    a 3, 5, 7, 9, 11, …   **b** 5, 9, 13, 17, 21, …   **c** 8, 13, 18, 23, 28, …

    d 2, 8, 14, 20, 26, …   **e** 5, 8, 11, 14, 17, …   **f** 2, 9, 16, 23, 30, …

3 For each sequence **a** to **f**, find:

    i the nth term   **ii** the 100th term   **iii** the term closest to 100.

        a 4, 7, 10, 13, 16, …   **b** 7, 9, 11, 13, 15, …   **c** 3, 8, 13, 18, 23, …

        d 1, 5, 9, 13, 17, …   **e** 2, 10, 18, 26, …   **f** 5, 6, 7, 8, 9, …

4 A sequence of fractions is $\frac{3}{5}, \frac{5}{8}, \frac{7}{11}, \frac{9}{14}, \dots$

    a Find the nth term in the sequence.

    b Change each fraction to a decimal.

    c What, as a decimal, will be the value of the:

       i 100th term   **ii** 1000th term?

    d Use your answers to part **c** to predict what the 10 000th term and the millionth term will be. (Check your predictions on a calculator.)

5 a A number pattern begins 1, 1, 2, 3, 5, 8, …

   i What is the next number in this pattern?

   ii The number pattern is continued. Explain how you would find the 10th number in the pattern.

b Another number pattern begins 1, 5, 9, 13, 17, …

   Write down the nth term in this pattern.

6 A taxi firm uses this chart for the charges for journeys of k kilometres.

k	1	2	3	4	5	6	7	8	9	10
Charge (£)	4.50	6.50	8.50	10.50	12.50	15.00	17.00	19.00	21.00	23.00
k	11	12	13	14	15	16	17	18	19	20
Charge (£)	26.00	28.00	30.00	32.00	34.00	37.00	39.00	41.00	43.00	45.00

a Using the charges for 1 to 5 km, work out an expression for the kth term.

b Using the charges for 6 to 10 km, work out an expression for the kth term.

c Using the charges for 10 to 15 km, work out an expression for the kth term.

d Using the charges for 16 to 20 km, work out an expression for the kth term.

e What is the basic charge per kilometre?

7 A series of fractions is $\frac{3}{7}, \frac{5}{10}, \frac{7}{13}, \frac{9}{16}, \frac{11}{19}$ …

a Write down the nth term of the numerators.

b Write down the nth term of the denominators.

c i Work out the fraction when $n = 1000$.

   ii Give the answer as a decimal.

d Will the terms of the series ever be greater than $\frac{2}{3}$? Give reasons for your answer.

4.4 Special sequences

Homework 4D

1 p is an even number, q is a square number.

State whether each expression is always odd, always even or could be either odd or even.

a $p + 1$ **b** $p + q$ **c** $2q$ **d** $3p - 1$

e p^2 **f** $pq - 1$ **g** $p^2 + 4q$ **h** $(p + q)(p - q)$

2 The powers of 3 are $3^1, 3^2, 3^3, 3^4$, …

This gives the sequence 3, 9, 27, 81, …

a Continue the sequence for another three terms.

The nth term of the powers of 3 is given by 3^n.

b Give the nth term of these sequences.

   i 2, 8, 26, 80, …   **ii** 6, 18, 54, 162, …

3 The negative powers of 10, starting with the power 0, are 10^0, 10^{-1}, 10^{-2}, 10^{-3}, ...

This gives the sequence 1, 0.1, 0.01, 0.001, …

  a Describe the connection between the number of zeros after the decimal point and before the 1, and the power of the term.

  b If $10^{-n} = 0.0000001$, what is the value of n?

4 **a** Copy and complete the table to show whether the result of adding prime, odd and even numbers, is always odd, always even or could be either odd or even.

+	Prime	Odd	Even
Prime	Either		
Odd		Even	
Even			Even

  b Copy and complete the table to show the whether the result of multiplying prime, odd and even numbers is always odd, always even or could be either odd or even.

×	Prime	Odd	Even
Prime	Either		
Odd		Odd	
Even			Even

5 **a** Draw an equilateral triangle with side length 9 cm on centimetre triangular paper.

    Work out the perimeter of this shape.

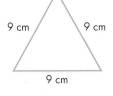

  b Divide each side of your shape from part **a** into three equal parts. On each side draw an equilateral triangle with side length 3 cm, each time using the centre section as one side.

    Work out the perimeter of the new shape.

  c Divide each side of your shape from part **b** into three equal parts. On each side draw an equilateral triangle with side length 1 cm, each time using the centre section as one side.

    Work out the perimeter of the new shape.

  d The next step would be to draw a triangle with side length $\frac{1}{3}$ cm from the centre section of each side of your shape from part **c**, but it becomes increasingly difficult to draw the shapes accurately.

    Look at the pattern of the perimeters of the first three shapes and write down the perimeter of the fourth shape.

  e The formula for the pattern of the perimeters is $27 \times \left(\frac{4}{3}\right)^{n-1}$.

    Use a calculator to check your perimeters for $n = 1 \ldots n = 4$.

    Work out the perimeter when $n = 100$

    If you keep on drawing triangles, the perimeter would become infinite.

    This is an example of a shape with a finite area and an infinite perimeter.

4.5 General rules from given patterns

Homework 4E

1 This pattern of shapes is built up from matchsticks.

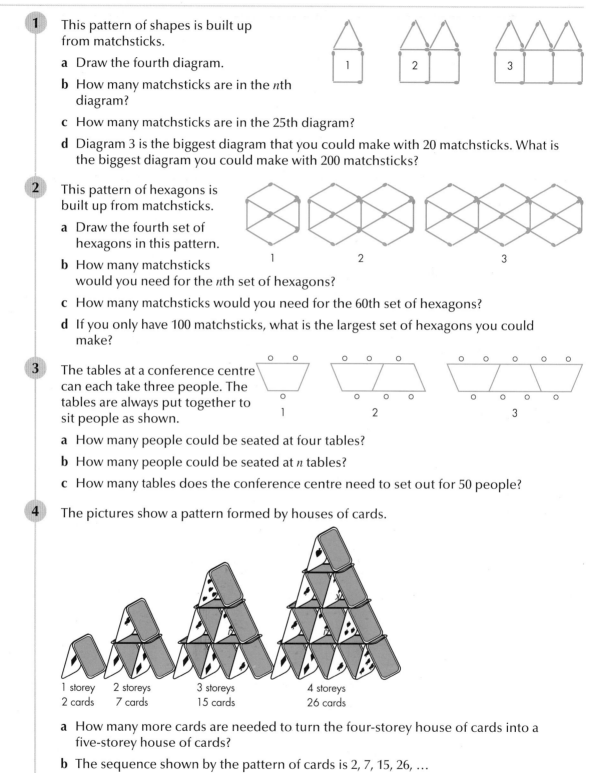

  a Draw the fourth diagram.

  b How many matchsticks are in the nth diagram?

  c How many matchsticks are in the 25th diagram?

  d Diagram 3 is the biggest diagram that you could make with 20 matchsticks. What is the biggest diagram you could make with 200 matchsticks?

2 This pattern of hexagons is built up from matchsticks.

  a Draw the fourth set of hexagons in this pattern.

  b How many matchsticks would you need for the nth set of hexagons?

  c How many matchsticks would you need for the 60th set of hexagons?

  d If you only have 100 matchsticks, what is the largest set of hexagons you could make?

3 The tables at a conference centre can each take three people. The tables are always put together to sit people as shown.

  a How many people could be seated at four tables?

  b How many people could be seated at n tables?

  c How many tables does the conference centre need to set out for 50 people?

4 The pictures show a pattern formed by houses of cards.

1 storey   2 storeys   3 storeys   4 storeys
2 cards   7 cards   15 cards   26 cards

  a How many more cards are needed to turn the four-storey house of cards into a five-storey house of cards?

  b The sequence shown by the pattern of cards is 2, 7, 15, 26, …

   i What is the sixth term in this sequence?

   ii Explain how you found your answer.

5 **a** Find the first three terms that these two sequences have in common:

2, 5, 8, 11, 14, ...

3, 7, 11, 15, 19, ...

b Write down the nth term of the sequence that is the answer to part **a**.

4.6 and 4.7 (Finding) The nth term of a quadratic sequence

Homework 4F

1 Write down the next two terms of each sequence and say how it is building up.

a 10, 13, 18, 25, 34, … , …   **b** 5, 8, 13, 20, 29, … , …

c −5, −2, 3, 10, 19, … , …   **d** 3, 6, 11, 18, 27, … , …

e −2, 1, 6, 13, 22, … , …

2 Write down the first five terms of the sequence for each nth term.

a n^2 **b** $n^2 - 1$ **c** $2n^2 + 2$ **d** $3n^2 - 3$ **e** $2n^2 - n - 1$

3 **a** Write down the first five terms of the quadratic sequence with the nth term $3n^2 - n - 2$.

b Find the first and second differences for this sequence.

c What do you notice about these differences?

4 Find the nth term of each quadratic sequence.

a 1, 7, 17, 31, 49   **b** 5, 8, 13, 20, 29   **c** 4, 13, 28, 49, 76

d 7, 13, 23, 37, 55   **e** 1, 13, 33, 61, 97

5 This pattern is made from black and white tiles.

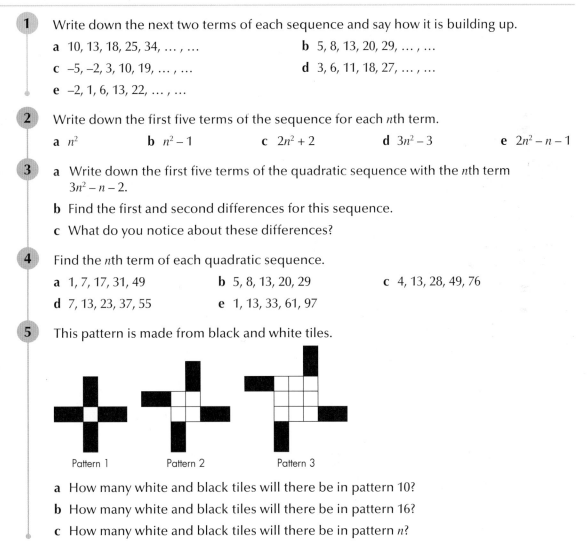

Pattern 1   Pattern 2   Pattern 3

a How many white and black tiles will there be in pattern 10?

b How many white and black tiles will there be in pattern 16?

c How many white and black tiles will there be in pattern n?

5 Ratio and proportion and rates of change: Ratio and proportion

5.1 Ratio

Homework 5A

1 Express each ratio in its simplest form.

    a $3:9$ **b** $5:25$ **c** $4:24$ **d** $10:30$ **e** $6:9$

    f $12:20$ **g** $25:40$ **h** $30:4$ **i** $14:35$ **j** $125:50$

2 Write each ratio of quantities in its simplest form. (Remember to change to a common unit first, where necessary.)

    a £2 to £8   **b** £12 to £16   **c** 25 g to 200 g

    d 6 miles : 15 miles   **e** 20 cm : 50 cm   **f** 80p : £1.50

    g 1 kg : 300 g   **h** 40 seconds : 2 minutes   **i** 9 hours : 1 day

3 Bob and Kathryn share £20 in the ratio $1:3$.

    a What fraction of the £20 does Bob receive?

    b What fraction of the £20 does Kathryn receive?

4 In a class of students, the ratio of boys to girls is $2:3$.

    a What fraction of the class are boys?   **b** What fraction of the class are girls?

5 Pewter is an alloy containing lead and tin in the ratio $1:9$.

    a What fraction of pewter is lead?   **b** What fraction of pewter is tin?

6 Manuel wins two-thirds of his snooker matches. He loses the rest.

    What is his ratio of wins to losses?

7 In the 2013 Ashes cricket series, the numbers of wickets taken by Chris Broad and Graham Swann were in the ratio $5:1$.

    The ratio of the number of wickets taken by James Anderson to those taken by Chris Broad was $2:1$.

    What fraction of the wickets taken by these three bowlers was taken by Graham Swann?

8 In a motor-sales business, the ratio of the area allocated to selling new cars to the area for selling used cars is $3:2$.

    Half the area for selling new cars is changed to used cars.

    What is the ratio of the areas now? Give your answer in its simplest form.

Homework 5B 🖩

1. Divide each amount according to the given ratio.
   a £10 in the ratio 1 : 4   b £12 in the ratio 1 : 2   c £40 in the ratio 1 : 3
   d 60 g in the ratio 1 : 5   e 10 hours in the ratio 1 : 9

2. The ratio of female to male members at a sports centre is 3 : 1. The total number of members of the centre is 400.
   a How many members are female?   b How many members are male?

3. A 20-metre length of cloth is cut into two pieces in the ratio 1 : 9. How long is each piece?

4. James collects coins. The ratio of silver coins to bronze coins in his collection is 5 : 2. He has 1400 coins. How many bronze coins does he have?

5. Maurice and Jane share a box of sweets in the ratio of their ages. Maurice is 9 years old and Jane is 11 years old. If there are 100 sweets in the box, how many does Maurice get?

6. Emily is given £30 for her birthday. She decides to spend four times as much as she saves. How much does she save?

7. Mrs Megson calculates that her quarterly gas and electricity bills are in the ratio 6 : 5. The total she pays for both bills is £66. How much is each bill?

8. You can simplify a ratio by changing it into the form 1 : n. For example, you can rewrite 5 : 7 as 1 : 1.4 by dividing each side of the ratio by 5. Rewrite each ratio in the form 1 : n.
   a 2 : 3   b 2 : 5   c 4 : 5   d 5 : 8   e 10 : 21

9. The amount of petrol and diesel sold at a garage is in the ratio 2 : 1. One-tenth of the diesel sold is bio-diesel.
   What fraction of all the fuel sold is bio-diesel?

10. At a buffet, there are twice as many men as women.
    The total money taken is £600.
    There are 50 men at the buffet.
    How much does each person pay for the buffet?

11. The cost of show tickets for adults and children is in the ratio 5 : 3.
    30 adults and 40 children visit the show.
    Children's tickets cost £3.60.
    How much money will the show take from these ticket sales?

Homework 5C 🖩

1. Peter and Margaret's ages are in the ratio 4 : 5. If Peter is 16 years old, how old is Margaret?

2. Cans of lemonade and packets of crisps were bought in the ratio 3 : 2 for a school disco. The organiser bought 120 cans of lemonade. How many packets of crisps did she buy?

3 Manuel is making fruit punch from fruit juice and iced soda water in the ratio 2 : 3. Manuel uses 10 litres of fruit juice.

    a How many litres of soda water does he use?

    b How many litres of fruit punch does he make?

4 Cupro-nickel coins are minted by mixing copper and nickel in the ratio 4 : 1.

    a How much copper is needed to mix with 20 kg of nickel?

    b How much nickel is needed to mix with 20 kg of copper?

5 The ratio of male to female spectators at a school inter-form football match is 2 : 1.

    If 60 boys watched the game, how many spectators were there in total?

6 Marmalade is made from sugar and oranges in the ratio 3 : 5. A jar of 'Savilles' marmalade contains 120 g of sugar.

    a What is the mass of oranges in the jar?

    b What is the total mass of the marmalade in the jar?

7 Fred's blackcurrant juice is made from 4 parts blackcurrant and 1 part water.

    Jodie's blackcurrant juice is made from blackcurrant and water in the ratio 7 : 2.

    Which juice contains the greater proportion of blackcurrant?

    Show how you work out your answer.

8 Sand and cement is mixed in the ratio 3 : 1.

    Cement is sold in 25-kg bags.

    Sand is sold in 875-kg sacks.

    How many sacks of sand are needed to mix with 20 bags of cement?

9 The ratio of tins of white paint to coloured paint in a shop storeroom is 2 : 5.

    There is enough room on the shelves for 60 tins of paint.

    How many tins of white paint can be put on the shelf if the ratio of the tins of white paint to coloured paint is also 2 : 5?

5.2 Direct proportion problems

Homework 5D

1 If four DVDs cost £3.20, what will 10 DVDs cost?

2 Five oranges cost 90p. Find the cost of 12 oranges.

3 Dylan earns £18.60 in 3 hours. How much will he earn in 8 hours?

4 Barbara bought 12 postcards for 3 euros when she was on holiday in Tenerife.

    a How many euros would she have paid for 9 postcards?

    b How many postcards could she have bought for 5 euros?

5. Five 'Day-Rover' bus tickets cost £8.50.

    a How much will 16 tickets cost?

    b Pat has £20. She wants to buy 12 'Day-Rover' bus tickets.

       Does she have enough money? Show your working.

6. A car uses 8 litres of petrol on a trip of 72 miles.

    a How much would the same car use on a trip of 54 miles?

    b What distance would the car travel on a full tank of 45 litres of petrol?

7. It takes a photocopier 18 seconds to produce 12 copies. How long will it take to produce 32 copies?

8. Val has a recipe for making 12 flapjacks.

    100 g margarine   4 tablespoons golden syrup

    80 g granulated sugar   200 g rolled oats

    a What quantities are need for:

        i 6 flapjacks   ii 24 flapjacks   iii 30 flapjacks?

    b What is the maximum number of flapjacks she can make if she has 1 kg of each ingredient?

9. Greg the baker sells bread rolls in pack of 6 for £1.

    Dom the baker sells bread rolls in packs of 24 for £3.19.

    I have £5 to spend on bread rolls.

    How many more rolls can I buy from Greg than from Dom?

10. To make white coffee, one-quarter of a cup is filled with milk.

    A cup holds 600 ml of white coffee altogether.

    How many cups of coffee can be made with 1 litre of milk?

11. A nurse can examine 20 patients each hour.

    There are 170 patients visiting a three-hour clinic.

    How many nurses are needed?

5.3 Best buys

Homework 5E

1. State which price in each pair, if either, is the better buy.

    a Mouthwash: £1.99 for a twin pack, £1.49 each with a 3 for 2 offer

    b Dusters: 79p for a pack of 6 with a 'buy one pack and get one pack free' offer, £1.20 for a pack of 20

2. State which price in each pair is the better buy. Give reasons for your choice.

    a Tomato ketchup: a medium bottle (200 g) for 55p, a large bottle (350 g) for 87p

    b Milk chocolate: a small bar (125 g) for 77p, a large bar (200 g) for 92p

    c Coffee: a large tin (750 g) for £11.95, a small tin (500 g) for £7.85

    d Honey: a large jar (900 g) for £2.35, a small jar (225 g) for 65p

3 Boxes of 'Wetherels' teabags are sold in three different sizes.

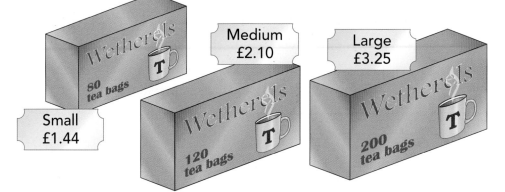

Which size of box gives the best value for money?

4 Bottles of Cola are sold in different sizes.

a Copy and complete the table.

Size of bottle	Price	Cost per litre
$\frac{1}{2}$ litre	36p	
$1\frac{1}{2}$ litres	99p	
2 litres	£1.40	
3 litres	£1.95	

b Which size of bottle gives the best value for money?

5 These products were being promoted by a supermarket.

Which offer is better value for money? Explain why.

6 Hannah scored 17 out of 20 in a test. John scored 40 out of 50 in a test of the same standard.

Who got the better mark?

5.4 Compound measures

Homework 5F

1 Work out the total pay for each person.

a 50 hours at £6 per hour

b $40\frac{1}{2}$ hours at £4.20 per hour

c 25 hours at £9.50 per hour

d $85\frac{1}{2}$ hours at £120 per hour

2 Work out the hourly rate for each payment.

a £450 for 50 hours' work

b £280 for 35 hours' work

c £207 for 12 hours' work

d £1268.75 for 80 hours' work

3 Work out the number of hours worked for each job.

   a £254 at £6.35 per hour   **b** £612 at £10.20 per hour

   c £362.25 at £8.05 per hour   **d** £816 at £25.50 per hour

4 Jane's normal working week is from 8 am to 4 pm, Monday to Friday, with a 1-hour lunch break. Her hourly rate is £20.50.

If she works Saturday or Sunday, she is paid 'double time', which means she gets twice her normal hourly rate.

   a How much does Jane earn for a normal working week?

   b One week she works for 6 hours on Saturday. How much does she earn that week?

5 Han works for 45 hours at his normal hourly rate and 12 hours at 'time and a half' which means he gets one and a half times his normal rate. He earns a total of £535.50. What is his hourly rate of pay?

6 One week Kate works her normal hours and 10 hours at 'double time'. She is paid £360. The next week she works her normal hours and 6 hours at 'time and a half'. She is paid £294. Work out how many hours she works in a normal week and her hourly pay rate?

Homework 5G

1 A cyclist travels a distance of 60 miles in 4 hours. What was his average speed?

2 How far along a motorway would you travel if you drove at an average speed of 60 mph for 3 hours?

3 Mr Baylis drives from Manchester to London in $4\frac{1}{2}$ hours. The distance is 207 miles. What is his average speed?

4 The distance from Leeds to Birmingham is 125 miles. The train I catch travels at an average speed of 50 mph. If I catch the 11.30 am train from Leeds, at what time should I expect to arrive in Birmingham?

5 Copy and complete this table.

	Distance travelled	Time taken	Average speed
a	240 miles	8 hours	
b	150 km	3 hours	
c		4 hours	5 mph
d		$2\frac{1}{2}$ hours	20 km/h
e	1300 miles		400 mph
f	90 km		25 km/h

6 A coach travels at an average speed of 60 km/h for 2 hours on a motorway, then slows down to do the last 30 minutes of its journey at an average speed of 20 km/h.

   a What is the total distance of this journey?

   b What is the average speed of the coach over the whole journey?

7 Hilary cycles six miles to work each day. She cycles the first 5 miles at an average speed of 15 mph and then the last mile in 10 minutes.

    a How long does it take Hilary to get to work?

    b What is her average speed for the whole journey?

8 Martha drives home from work in 1 hour 15 minutes. She drives home at an average speed of 36 mph.

    a Change 1 hour 15 minutes to decimal time in hours.

    b How far is it from Martha's work to her home?

9 A tram from A to B takes 15 minutes at an average speed of 16 mph.

    The distance from A to B by car is two miles longer.

    How fast would a car need to travel to get from A to B in the same time as the tram?

10 A taxi travelled for 30 minutes. The fare was £24.

    If the fare was charged at £1.20 per mile, what was the average speed of the taxi?

11 Two cars are currently 30 miles apart, but travelling towards each other.

    The average speed of one car is twice as fast as the other car.

    The slower car is averaging 20 mph.

    How long will it be before they meet up?

Homework 5H 🖩

1 Find the density of a piece of wood with a mass of 135 g and a volume of 150 cm³.

2 Calculate the density of a metal if 40 cm³ of it has a mass of 2500 g.

3 A force of 40 N acts over an area of 10 m². What is the pressure?

4 A pressure of 8 Pa acts on an area of $\frac{1}{4}$ m². What force is exerted?

5 Calculate the mass of a piece of plastic, 25 cm³ in volume, if its density is 1.2 g/cm³.

6 Calculate the volume of a piece of wood which has a mass of 350 g and a density of 0.7 g/cm³.

7 Find the mass of a marble statue with a volume of 540 cm³, if the density of marble is 2.5 g/cm³.

8 Calculate the volume of a liquid with a mass of 1 kg and a density of 1.1 g/cm³.

9 Find the density of a stone with a mass of 63 g and a volume of 12 cm³.

10 It is estimated that a huge rock balanced on the top of a mountain has a volume of 120 m³. The density of the rock is 8.3 g/cm³. What is the estimated mass of the rock?

11 A 1-kg bag of flour has a volume of about 900 cm³. What is the density of flour in g/cm³?

12 $1 \text{ m}^3 = 1\,000\,000 \text{ cm}^3$

A storage area has 30 tonnes of sandstone.

The density of the sandstone is 2.3 g/cm³.

a What is the volume of the sandstone in the storage area in m³?

b The density of granite is 2.7 g/cm³.

The volume of granite stored is the same as the volume of sandstone.

How much heavier is the granite?

13 The density of a piece of oak is 630 kg/m³.

The density of a piece of mahogany is 550 kg/m³.

Two identical carvings are made, one from oak and the other from mahogany.

The oak carving has a mass of 315 g.

What is the mass of the mahogany carving?

14 Two metal objects appear to be identical but have different masses.

How does this tell you that they are probably made from different metals?

15 A small bronze statue has a mass of 50 kg.

When placed on its base, the pressure exerted is 10 000 Pa.

Work out the area of the base. (Take $g = 10 \text{ m/s}^2$)

5.5 Compound interest and repeated percentage change

Homework 5I

1 A small plant increases its height by 10% each day in the second week of its growth. At the end of the first week, the plant was 5 cm high.

What is its height after a further:

a 1 day   **b** 2 days   **c** 4 days   **d** 1 week?

2 The headmaster of a new school offered his staff an annual pay increase of 5% for every year they stayed with the school.

a Mr Speed started teaching at the school on a salary of £28 000. What salary will he be on after 3 years?

b Miss Tuck started teaching at the school on a salary of £14 500. How many years will it be until she is earning a salary of over £20 000?

3 Billy put a gift of £250 into a special savings account that offered him 8% compound interest if he promised to keep the money in the account for at least 2 years. How much was in this account after:

a 2 years   **b** 3 years   **c** 5 years?

4 The penguin population of a small island was only 1500 in 2008, but increased steadily by about 15% each year. Calculate the population in:

a 2009   **b** 2010   **c** 2012.

5 A sycamore tree is 40 cm tall. It grows at a rate of 8% per year. A conifer is 20 cm tall. It grows at a rate of 15% per year. How many years does it take before the conifer is taller than the sycamore?

6 The population of a small town is 2000. It is falling by 10% each year.

The population of a nearby village is 1500. It is rising by 10% each year.

After how many years will the population of the town be less than the population of the village?

7 Each week, a boy takes out 20% of the amount in his bank account to spend.

After how many weeks will the amount in his bank account have halved from the original amount?

5.6 Reverse percentage (working out the original amount)

Homework 5J

1 Find what 100% represents in each situation.

a 20% represents 160 g   **b** 25% represents 24 m   **c** 5% represents 42 cm

2 Find what 100% represents in each situation.

a 40% represents 28 kg   **b** 30% represents £54   **c** 15% represents 6 hours

3 VAT is a government tax added to goods and services. With VAT at 17.5%, what is the pre-VAT price of these goods?

Jumper £14.10   Socks £1.88   Trousers £23.50

4 Paula spends £9 each week on CDs. This is 60% of her weekly allowance. How much is Paula's weekly allowance?

5 Yuni's weekly pay is increased by 4% to £187.20. What was Yuni's pay before the increase?

6 Jon's salary is £23 980. This is 10% more than he earned two years ago.

Last year his salary was 3% more than it was two years ago.

a How much was his salary last year?

b By what percentage did he salary increase last year?

7 Twice as many people visit a shopping centre on Saturdays as on Fridays.

The numbers visiting on both days increases by 50% in the week before Christmas.

How many more visit on this Saturday than on this Friday? Give your answer as a percentage.

8 A man's savings decreased by 10% in one year and then increased in the following year by 10%. He now has £1782.

How much did he have two years ago?

6 Geometry and measures: Angles

6.1 Angle facts

Homework 6A 🖩

1 Calculate the size of the angle marked x in each diagram.

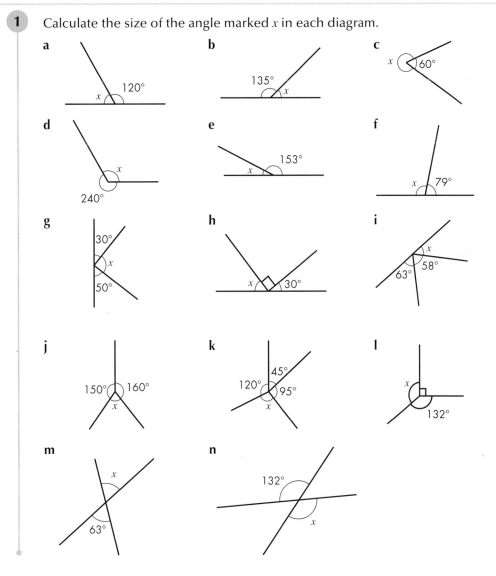

2 Will these three angles fit together to make a straight line?

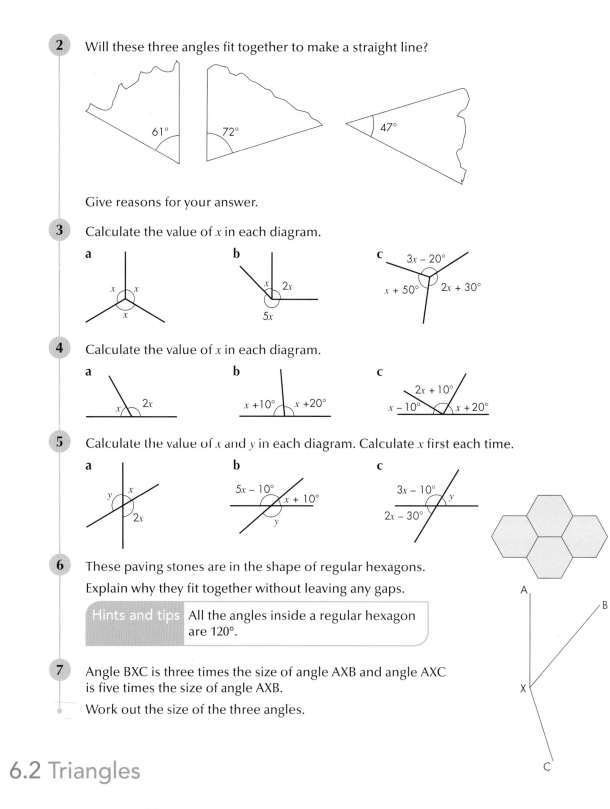

Give reasons for your answer.

3 Calculate the value of x in each diagram.

a

b

c
$3x - 20°$
$x + 50°$ $2x + 30°$

4 Calculate the value of x in each diagram.

a
$2x$
x

b
$x + 10°$ $x + 20°$

c
$2x + 10°$
$x - 10°$ $x + 20°$

5 Calculate the value of x and y in each diagram. Calculate x first each time.

a
y x
$2x$

b
$5x - 10°$
$x + 10°$
y

c
$3x - 10°$
y
$2x - 30°$

6 These paving stones are in the shape of regular hexagons.

Explain why they fit together without leaving any gaps.

> **Hints and tips** All the angles inside a regular hexagon are 120°.

7 Angle BXC is three times the size of angle AXB and angle AXC is five times the size of angle AXB.

Work out the size of the three angles.

6.2 Triangles

Homework 6B 🖩

1 Each set of angles form the three interior angles of a triangle. Find the value of the angle indicated by a letter in each set.

a 40°, 70° and $a°$ **b** 60°, 60° and $b°$ **c** 80°, 90° and $c°$

d 65°, 72° and $d°$ **e** 130°, 45° and $e°$ **f** 112°, 27° and $f°$

2 State whether each set of angles form the three interior angles of a triangle? Give reasons for your answers.

    a 15°, 85° and 80°   **b** 40°, 60° and 90°   **c** 25°, 25° and 110°

    d 40°, 40° and 100°   **e** 32°, 37° and 111°   **f** 61°, 59° and 70°

3 Find the size of the angle marked with a letter in each diagram.

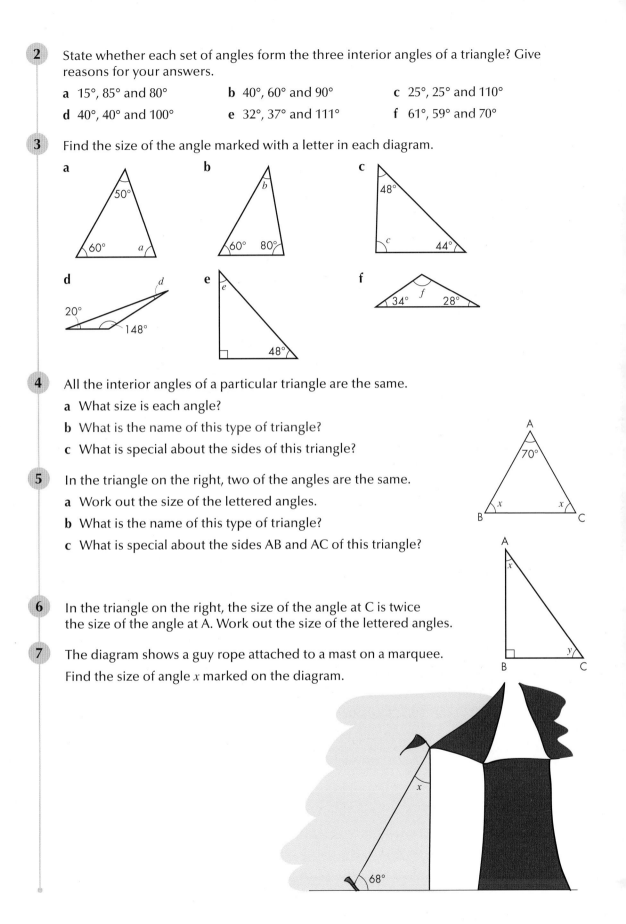

    a   **b**   **c**

    d   **e**   **f**

4 All the interior angles of a particular triangle are the same.

    a What size is each angle?

    b What is the name of this type of triangle?

    c What is special about the sides of this triangle?

5 In the triangle on the right, two of the angles are the same.

    a Work out the size of the lettered angles.

    b What is the name of this type of triangle?

    c What is special about the sides AB and AC of this triangle?

6 In the triangle on the right, the size of the angle at C is twice the size of the angle at A. Work out the size of the lettered angles.

7 The diagram shows a guy rope attached to a mast on a marquee.

    Find the size of angle *x* marked on the diagram.

8 Find the size of the angle marked with a letter in each diagram.

a

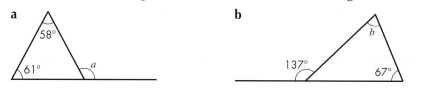

b

9 Here are five statements about triangles. Some are true and some are false.

A A triangle can have three acute angles.

B A triangle can have two acute angles and one obtuse angle.

C A triangle can have one acute angle and two obtuse angles.

D A triangle can have two acute angles and one right-angle.

E A triangle can have one acute angle and two right-angles.

   If a statement is true, draw a sketch of a triangle. If a statement is false, explain why.

10 Show that angle a is 25 degrees.

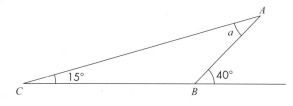

11 A town planner has drawn this diagram to show three paths in a park, but they have missed out the angle marked x.

Work out the value of x.

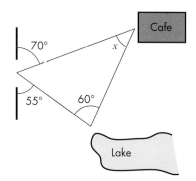

6.3 Angles in a polygon

Homework 6C

1 **a** Draw a diagram to show why the sum of the interior angles of any pentagon is 540°.

   b Find the size of the angle x in the pentagon.

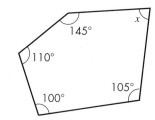

2 Calculate the size of the angle marked with a letter in each polygons.

a

b

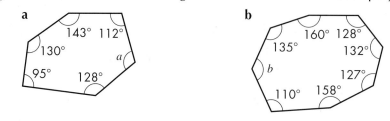

3 Jamal is cutting metal from a rectangular sheet to make this sign.

He needs to cut the two angles marked *x* accurately. What is the size of each one?

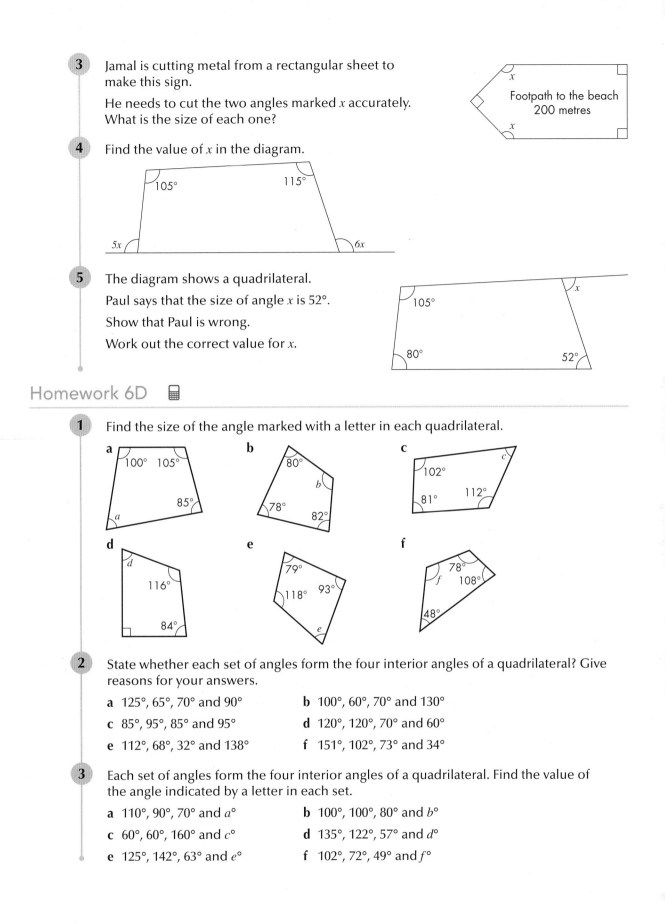

Footpath to the beach
200 metres

4 Find the value of *x* in the diagram.

105°   115°

5*x*   6*x*

5 The diagram shows a quadrilateral.

Paul says that the size of angle *x* is 52°.

Show that Paul is wrong.

Work out the correct value for *x*.

105°

x

80°   52°

Homework 6D 🖩

1 Find the size of the angle marked with a letter in each quadrilateral.

a
100° 105°
85°
a

b
80°
b
78°   82°

c
c
102°
81°   112°

d
d
116°
84°

e
79°
118° 93°
e

f
78°
f 108°
48°

2 State whether each set of angles form the four interior angles of a quadrilateral? Give reasons for your answers.

a 125°, 65°, 70° and 90°
b 100°, 60°, 70° and 130°
c 85°, 95°, 85° and 95°
d 120°, 120°, 70° and 60°
e 112°, 68°, 32° and 138°
f 151°, 102°, 73° and 34°

3 Each set of angles form the four interior angles of a quadrilateral. Find the value of the angle indicated by a letter in each set.

a 110°, 90°, 70° and *a*°
b 100°, 100°, 80° and *b*°
c 60°, 60°, 160° and *c*°
d 135°, 122°, 57° and *d*°
e 125°, 142°, 63° and *e*°
f 102°, 72°, 49° and *f*°

4 For the quadrilateral on the right:

   a Find the size of angle x.

   b What type of angle is x?

   c What is the special name of a quadrilateral like this?

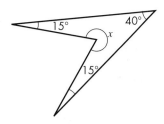

6.4 Regular polygons

Homework 6E

1 For each regular polygon below, find the size of the interior angle x and the exterior angle y.

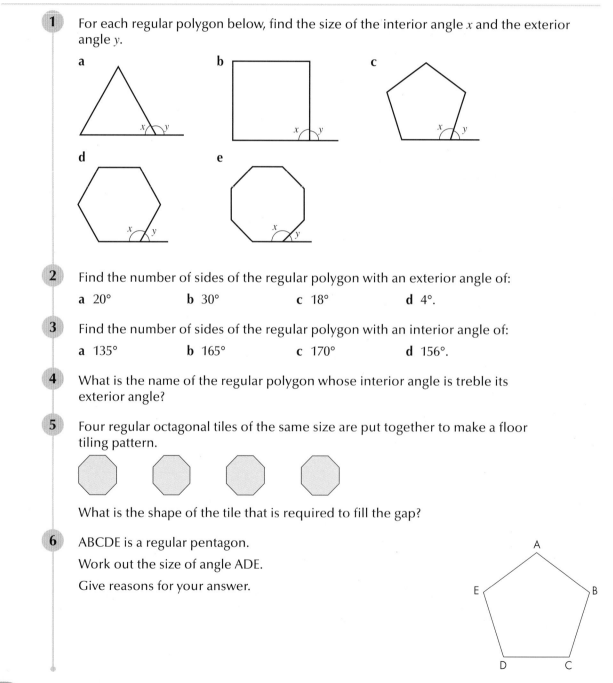

   a   **b**   **c**

   d   **e**

2 Find the number of sides of the regular polygon with an exterior angle of:

   a 20°   **b** 30°   **c** 18°   **d** 4°.

3 Find the number of sides of the regular polygon with an interior angle of:

   a 135°   **b** 165°   **c** 170°   **d** 156°.

4 What is the name of the regular polygon whose interior angle is treble its exterior angle?

5 Four regular octagonal tiles of the same size are put together to make a floor tiling pattern.

What is the shape of the tile that is required to fill the gap?

6 ABCDE is a regular pentagon.

Work out the size of angle ADE.

Give reasons for your answer.

7 Which of the following statements are true for a regular hexagon?

A The size of each interior angle is 60°.

B The size of each interior angle is 120°.

C The size of each exterior angle is 60°.

D The size of each exterior angle is 240°.

6.5 Angles in parallel lines

Homework 6F

1 State the size of the lettered angles in each diagram.

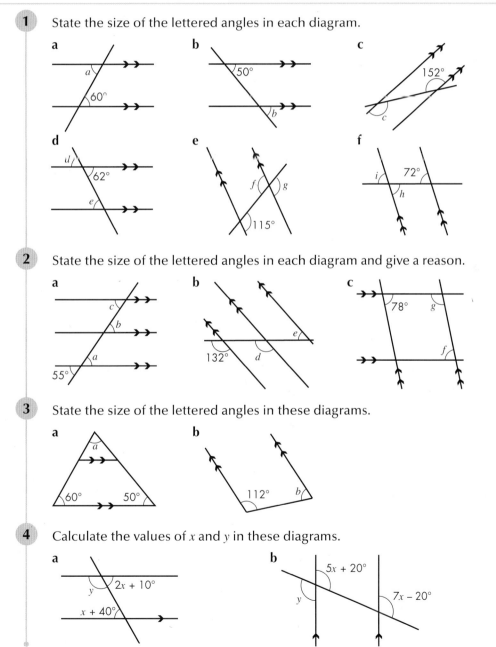

a

b

c

d

e

f

2 State the size of the lettered angles in each diagram and give a reason.

a

b

c

3 State the size of the lettered angles in these diagrams.

a

b

4 Calculate the values of x and y in these diagrams.

a

b

5 ABC is an isosceles triangle with angle ABC = 52°.

XY is parallel to BC.

Work out the size of angle BAC.

Explain clearly how you worked this out.

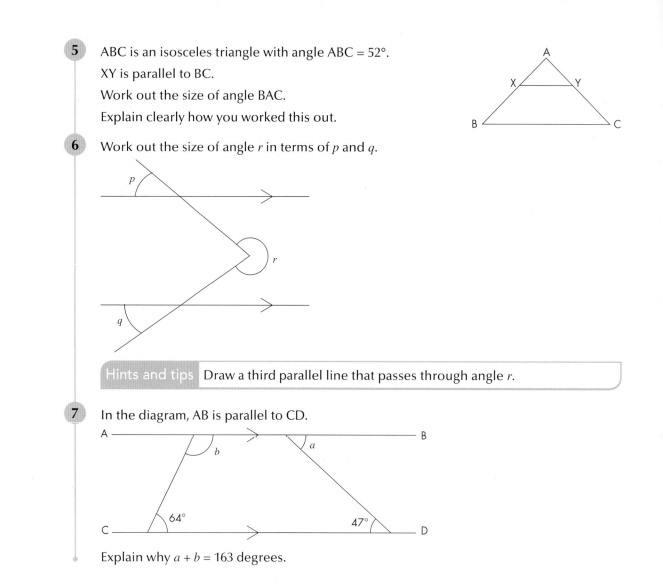

6 Work out the size of angle r in terms of p and q.

Hints and tips Draw a third parallel line that passes through angle r.

7 In the diagram, AB is parallel to CD.

Explain why $a + b = 163$ degrees.

6.6 Special quadrilaterals

Homework 6G

1 Calculate the sizes of the lettered angles in each trapezium.

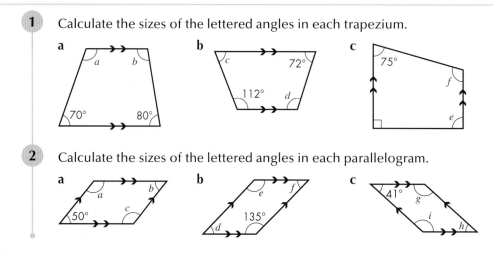

a

b

c

2 Calculate the sizes of the lettered angles in each parallelogram.

a

b

c

3 Calculate the sizes of the lettered angles in each kite

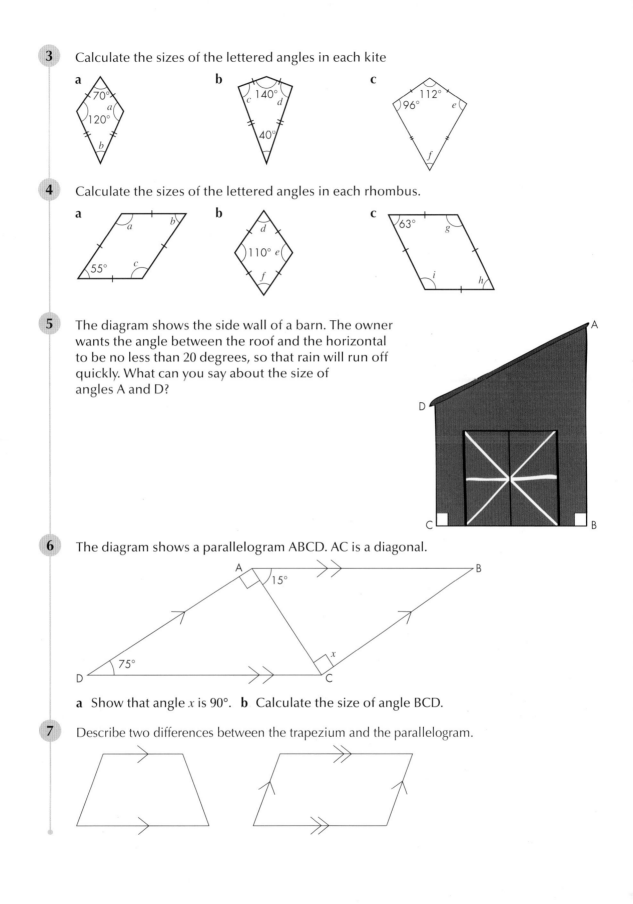

a
70°
a
120°
b

b
c 140° *d*
40°

c
112°
96° *e*
f

4 Calculate the sizes of the lettered angles in each rhombus.

a
a *b*
55° *c*

b
d
110° *e*
f

c
63° *g*
i *h*

5 The diagram shows the side wall of a barn. The owner wants the angle between the roof and the horizontal to be no less than 20 degrees, so that rain will run off quickly. What can you say about the size of angles A and D?

A
D
C B

6 The diagram shows a parallelogram ABCD. AC is a diagonal.

A 15° B
75° *x*
D C

a Show that angle *x* is 90°. **b** Calculate the size of angle BCD.

7 Describe two differences between the trapezium and the parallelogram.

Homework 6H 🖩

1. The diagram shows the floor plan of a kitchen. The scale is 1 cm to 30 cm.

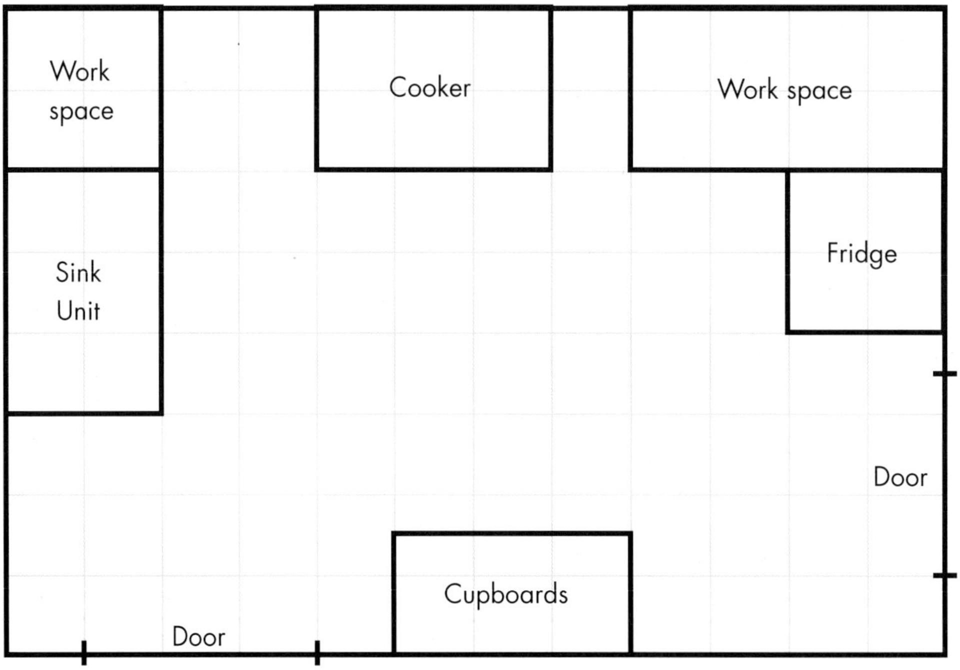

Work space		Cooker		Work space
Sink Unit				Fridge
				Door
		Cupboards		
Door				

   a State the actual dimensions of:

       i the sink unit   **ii** the cooker   **iii** the fridge   **iv** the cupboards.

   b Calculate the actual total area of work space.

2. The sketch shows a ladder leaning against a wall.

   The bottom of the ladder is 1 m away from the wall and it reaches 4 m up the wall.

   a Make a scale drawing to show the position of the ladder. Use a scale of 4 cm to 1 m.

   b Use your scale drawing to work out the actual length of the ladder.

3 The map below is drawn to a scale of 1 cm to 2 km.

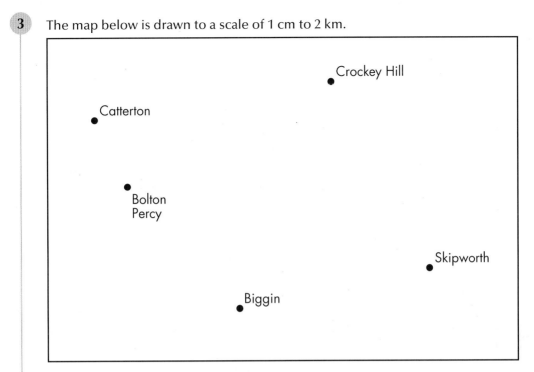

Give the approximate direct distances between:

a Biggin and Skipworth

b Bolton Percy and Crockey Hill

c Skipworth and Catterton

d Crockey Hill and Biggin

e Catterton and Bolton Percy.

4 This diagram shows a farmer's sketch of one of his fields.

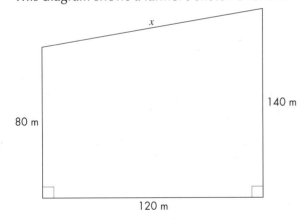

a Make a scale drawing of the field. Use the scale 1 cm represents 20 m.

b The farmer wants to build a wall along the side marked x on the diagram. Each metre length of wall uses 60 bricks.

Work out the number of bricks the farmer will need.

5 The map below shows the position of four fells in the Lake District. The map is drawn to a scale of 1 : 150 000.

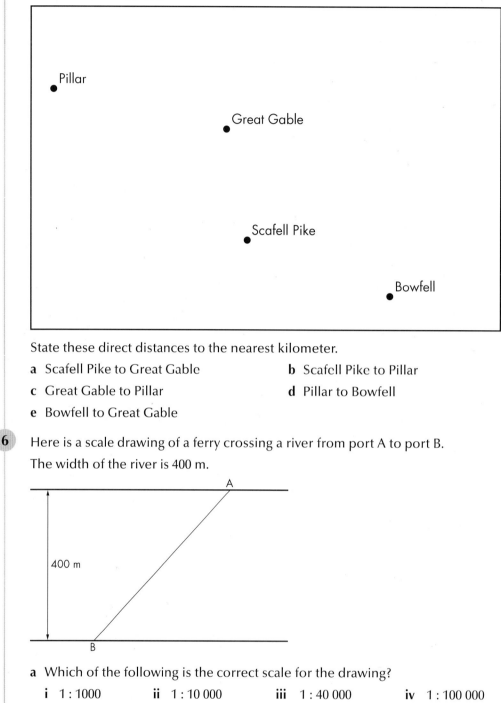

State these direct distances to the nearest kilometer.

a Scafell Pike to Great Gable   **b** Scafell Pike to Pillar

c Great Gable to Pillar   **d** Pillar to Bowfell

e Bowfell to Great Gable

6 Here is a scale drawing of a ferry crossing a river from port A to port B.
The width of the river is 400 m.

a Which of the following is the correct scale for the drawing?

  i 1 : 1000   **ii** 1 : 10 000   **iii** 1 : 40 000   **iv** 1 : 100 000

b What is the actual distance from port A to port B?

7 **a** Write down the bearing of B from A.   **b** Write down the bearing of D from C.

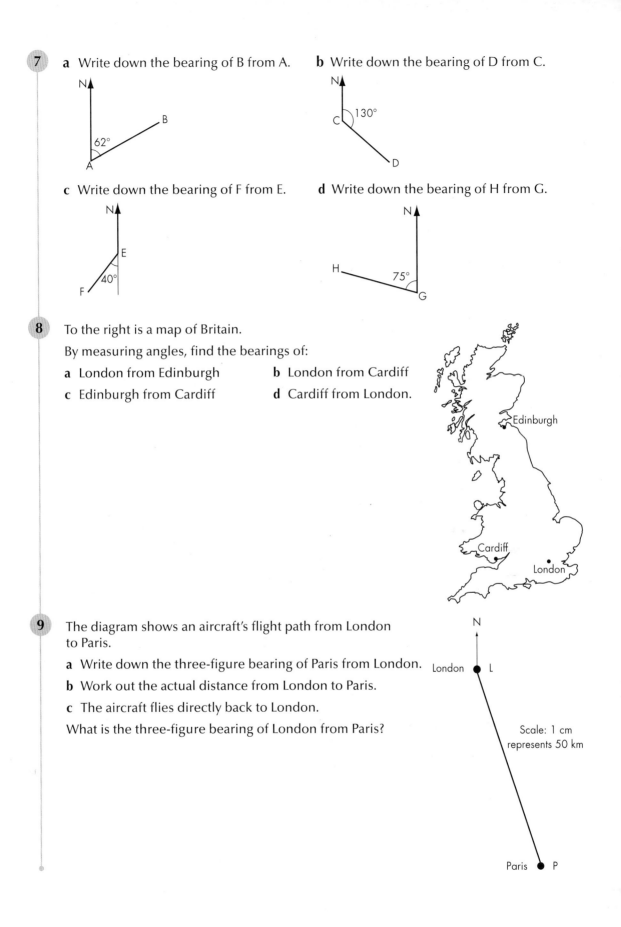

c Write down the bearing of F from E.   **d** Write down the bearing of H from G.

8 To the right is a map of Britain.

By measuring angles, find the bearings of:

a London from Edinburgh   **b** London from Cardiff

c Edinburgh from Cardiff   **d** Cardiff from London.

9 The diagram shows an aircraft's flight path from London to Paris.

a Write down the three-figure bearing of Paris from London.

b Work out the actual distance from London to Paris.

c The aircraft flies directly back to London.

What is the three-figure bearing of London from Paris?

Scale: 1 cm represents 50 km

10 **a** The bearing of B from A is $x°$.

What is the bearing of A from B?

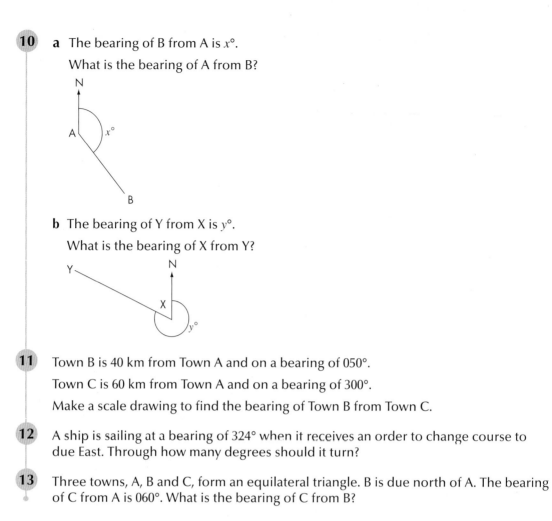

b The bearing of Y from X is $y°$.

What is the bearing of X from Y?

11 Town B is 40 km from Town A and on a bearing of 050°.

Town C is 60 km from Town A and on a bearing of 300°.

Make a scale drawing to find the bearing of Town B from Town C.

12 A ship is sailing at a bearing of 324° when it receives an order to change course to due East. Through how many degrees should it turn?

13 Three towns, A, B and C, form an equilateral triangle. B is due north of A. The bearing of C from A is 060°. What is the bearing of C from B?

7 Geometry and measures: Transformations, constructions and loci

7.1 Congruent triangles

Homework 7A

1 State whether each pair of triangles is congruent and, if so, the condition of congruency it satisfies.

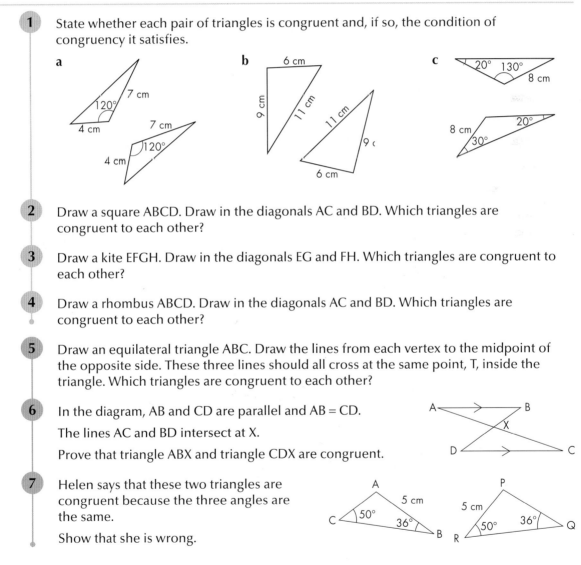

a

b

c

2 Draw a square ABCD. Draw in the diagonals AC and BD. Which triangles are congruent to each other?

3 Draw a kite EFGH. Draw in the diagonals EG and FH. Which triangles are congruent to each other?

4 Draw a rhombus ABCD. Draw in the diagonals AC and BD. Which triangles are congruent to each other?

5 Draw an equilateral triangle ABC. Draw the lines from each vertex to the midpoint of the opposite side. These three lines should all cross at the same point, T, inside the triangle. Which triangles are congruent to each other?

6 In the diagram, AB and CD are parallel and AB = CD.

The lines AC and BD intersect at X.

Prove that triangle ABX and triangle CDX are congruent.

7 Helen says that these two triangles are congruent because the three angles are the same.

Show that she is wrong.

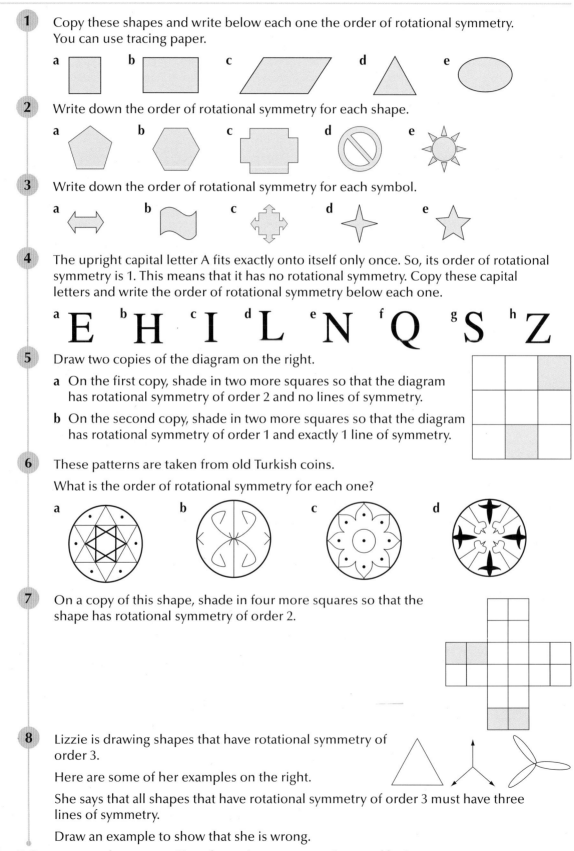

1 Copy these shapes and write below each one the order of rotational symmetry. You can use tracing paper.

   a   b   c   d   e

2 Write down the order of rotational symmetry for each shape.

   a   b   c   d   e

3 Write down the order of rotational symmetry for each symbol.

   a   b   c   d   e

4 The upright capital letter A fits exactly onto itself only once. So, its order of rotational symmetry is 1. This means that it has no rotational symmetry. Copy these capital letters and write the order of rotational symmetry below each one.

   a E   b H   c I   d L   e N   f Q   g S   h Z

5 Draw two copies of the diagram on the right.

   a On the first copy, shade in two more squares so that the diagram has rotational symmetry of order 2 and no lines of symmetry.

   b On the second copy, shade in two more squares so that the diagram has rotational symmetry of order 1 and exactly 1 line of symmetry.

6 These patterns are taken from old Turkish coins.

   What is the order of rotational symmetry for each one?

   a   b   c   d

7 On a copy of this shape, shade in four more squares so that the shape has rotational symmetry of order 2.

8 Lizzie is drawing shapes that have rotational symmetry of order 3.

   Here are some of her examples on the right.

   She says that all shapes that have rotational symmetry of order 3 must have three lines of symmetry.

   Draw an example to show that she is wrong.

7.3 Transformations

Homework 7C

1 Use vectors to describe these translations of the shapes on the grid.

   i A to B   **ii** A to C   **iii** A to D   **iv** B to A   **v** B to C   **vi** B to D

2 **a** Draw a set of coordinate axes with values of x and y from 0 to 10. Draw the triangle with coordinates A(4, 4), B(5, 7) and C(6, 5).

   b Draw the image of ABC after a translation with vector $\begin{pmatrix} 3 \\ 2 \end{pmatrix}$. Label this P.

   c Draw the image of ABC after a translation with vector $\begin{pmatrix} 4 \\ -3 \end{pmatrix}$. Label this Q.

   d Draw the image of ABC after a translation with vector $\begin{pmatrix} -4 \\ 3 \end{pmatrix}$. Label this R.

   e Draw the image of ABC after a translation with vector $\begin{pmatrix} -3 \\ -2 \end{pmatrix}$. Label this S.

3 Look at your diagram from question **2**. Describe the translation that will move:

   a P to Q   **b** Q to R   **c** R to S   **d** S to P

   e R to P   **f** S to Q   **g** R to Q   **h** P to S.

4 A group of hikers walk between three points A, B and C using direction vectors, with distances in kilometres.

The direction vector from A to B is $\begin{pmatrix} -4 \\ 3 \end{pmatrix}$ and the direction vector from B to C is $\begin{pmatrix} -2 \\ -5 \end{pmatrix}$.

   a Draw a diagram on centimetre-squared paper, to show the walk. Use a scale of 1 cm represents 1 km.

   b Work out the direction vector from C to A.

5   Write down a series of translations which will take you from the Start/finish, around the shaded square without touching it, and back to the Start/finish. Make as few translations as possible.

Start/finish

6   Joel says that if the translation from a point X to a point Y is described by the vector $\begin{pmatrix} -3 \\ 2 \end{pmatrix}$, then the translation from the point Y to the point X is described by the vector $\begin{pmatrix} 2 \\ -3 \end{pmatrix}$.

Is Joel correct? Show how you decide.

Homework 7D

1   a   Draw a pair of axes. Label the x-axis from –5 to 5 and the y-axis from –5 to 5.

   b   Draw the triangle with co-ordinates A(1, 1), B(5, 5) and C(3, 4).

   c   Reflect triangle ABC in the x-axis. Label the image P.

   d   Reflect triangle P in the y-axis. Label the image Q.

   e   Reflect triangle Q in the x-axis. Label the image R.

   f   Describe the reflection that will transform triangle ABC to triangle R.

2   Copy this diagram onto squared paper.

   a   Reflect triangle A in the line $x = 1$.
       Label the image B.

   b   Reflect triangle B in the line $y = -2$.
       Label the image C.

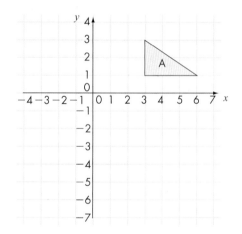

3   A designer is making a logo for a company.

   She starts with a kite ABCD.

   She then reflects the kite in the line BD on top of the original kite to obtain the logo.

   Draw any kite on squared paper to follow the designer's method to obtain a logo.

4   A point X has coordinates (a, b).

   Point X is reflected in the line $x = 3$

   Find the coordinates of the image of point X.

5 a Draw a pair of axes. Label the *x*-axis from –5 to 5 and the *y*-axis from –5 to 5.

  b Draw the triangle with co-ordinates A(2, 2), B(3, 4) and C(2, 4).

  c Reflect the triangle ABC in the line $y = x$. Label the image P.

  d Reflect the triangle P in the line $y = -x$. Label the image Q.

  e Reflect triangle Q in the line $y = x$. Label the image R.

  f Describe the reflection that will move triangle ABC to triangle R.

Homework 7E

1 Copy this diagram onto squared paper.

  a Rotate the shape 90° clockwise about (0, 0). Label the image P.

  b Rotate the shape 180° clockwise about (0, 0). Label the image Q.

  c Rotate the shape 90° anticlockwise about (0, 0). Label the image R.

  d What rotation takes R back to the original shape?

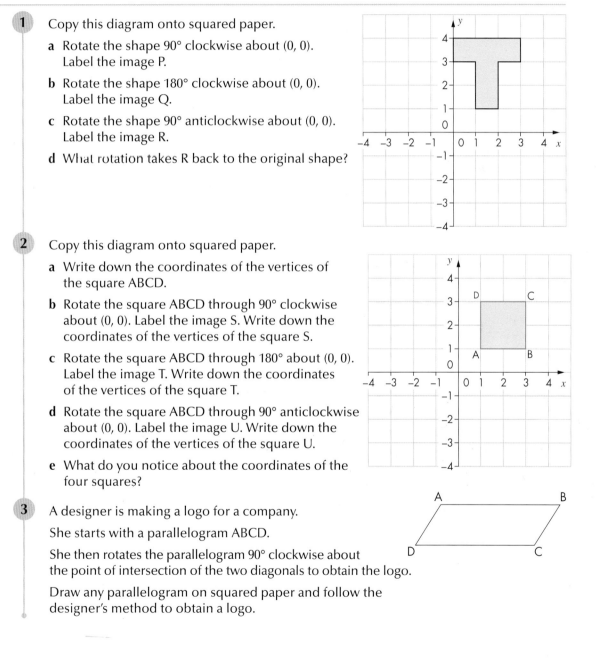

2 Copy this diagram onto squared paper.

  a Write down the coordinates of the vertices of the square ABCD.

  b Rotate the square ABCD through 90° clockwise about (0, 0). Label the image S. Write down the coordinates of the vertices of the square S.

  c Rotate the square ABCD through 180° about (0, 0). Label the image T. Write down the coordinates of the vertices of the square T.

  d Rotate the square ABCD through 90° anticlockwise about (0, 0). Label the image U. Write down the coordinates of the vertices of the square U.

  e What do you notice about the coordinates of the four squares?

3 A designer is making a logo for a company.

  She starts with a parallelogram ABCD.

  She then rotates the parallelogram 90° clockwise about the point of intersection of the two diagonals to obtain the logo.

  Draw any parallelogram on squared paper and follow the designer's method to obtain a logo.

4 Copy the diagram and rotate the given triangle as described.

a $\frac{1}{4}$ turn clockwise about (0, 0)

b $\frac{1}{2}$ turn clockwise about (0, 2)

c 90° anticlockwise about (−1, 1)

d 180° about (0, 0)

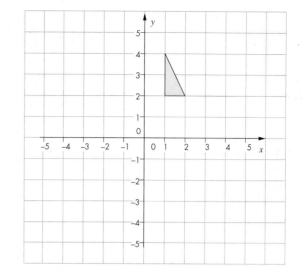

5 Describe the rotation that takes the shaded triangle to:

a triangle A

b triangle B

c triangle C

d triangle D.

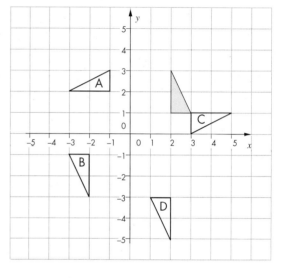

6 A point P has coordinates (a, b).

a The point P is rotated 90° clockwise about (0, 0) to give a point Q.

What are the coordinates of Q?

b The point P is rotated 180° clockwise about (0, 0) to give a point R.

What are the coordinates of R?

c The point P is rotated 90° anticlockwise about (0, 0) to give a point S.

What are the coordinates of S?

7 Triangle A, as shown on the grid, is rotated to form a new triangle B.

The coordinates of the vertices of B are (3, −1), (1, −4) and (3, −4).

Describe fully the rotation that maps triangle A onto triangle B.

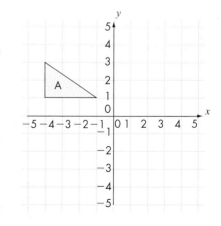

1   Copy each shape with its centre of enlargement. Use the ray method enlarge it by the given scale factor.

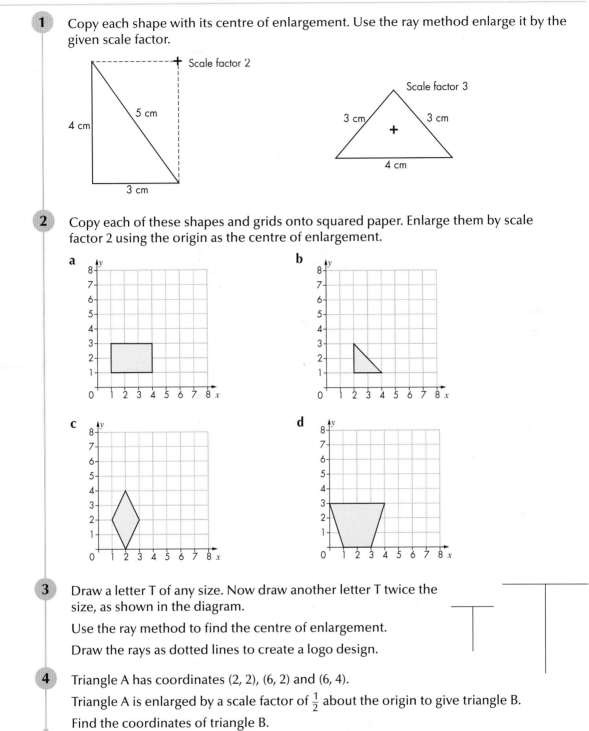

2   Copy each of these shapes and grids onto squared paper. Enlarge them by scale factor 2 using the origin as the centre of enlargement.

3   Draw a letter T of any size. Now draw another letter T twice the size, as shown in the diagram.

Use the ray method to find the centre of enlargement.

Draw the rays as dotted lines to create a logo design.

4   Triangle A has coordinates (2, 2), (6, 2) and (6, 4).

Triangle A is enlarged by a scale factor of $\frac{1}{2}$ about the origin to give triangle B.

Find the coordinates of triangle B.

5 Triangle B is an enlargement of triangle A.

Which of the following describes the enlargement?

a an enlargement of scale factor –2 about (0, 0)

b an enlargement of scale factor –3 about (0, 0)

c an enlargement of scale factor –3 about (1, 2)

d an enlargement of scale factor $-\frac{1}{3}$ about (1, 2)

Show how you decide.

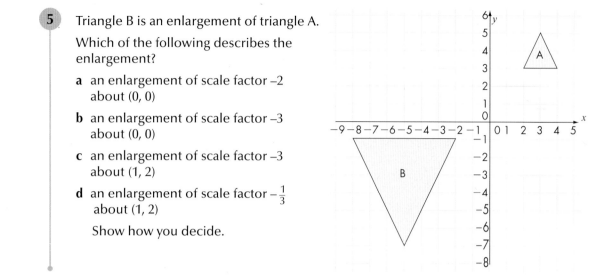

7.4 Combinations of transformations

Homework 7G

1 Describe fully the transformations that will result in the following movements.

a T_1 to T_2 **b** T_1 to T_6 **c** T_2 to T_3 **d** T_6 to T_2

e T_6 to T_5 **f** T_5 to T_4 **g** T_1 to T_5

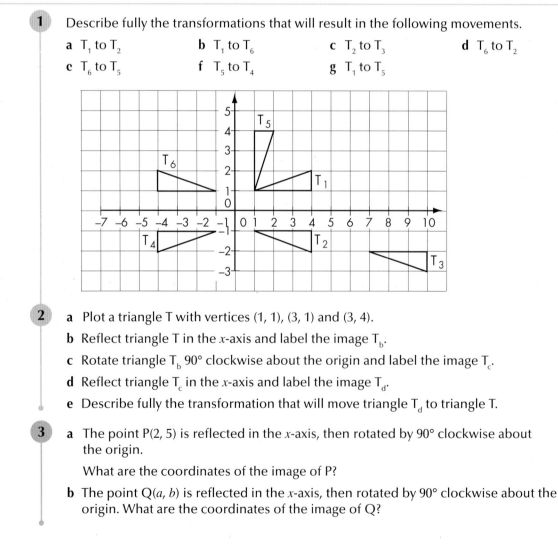

2 **a** Plot a triangle T with vertices (1, 1), (3, 1) and (3, 4).

b Reflect triangle T in the x-axis and label the image T_b.

c Rotate triangle T_b 90° clockwise about the origin and label the image T_c.

d Reflect triangle T_c in the x-axis and label the image T_d.

e Describe fully the transformation that will move triangle T_d to triangle T.

3 **a** The point P(2, 5) is reflected in the x-axis, then rotated by 90° clockwise about the origin.

What are the coordinates of the image of P?

b The point Q(a, b) is reflected in the x-axis, then rotated by 90° clockwise about the origin. What are the coordinates of the image of Q?

4   **a** The point R(4, 3) is reflected in the line $y = -x$, then reflected in the x-axis. What are the coordinates of the image of R?

     b The point S(a, b) is reflected in the line $y = -x$, then reflected in the x-axis. What are the coordinates of the image of S?

5   Copy the diagram onto squared paper.

     a Triangle A is translated by the

       vector $\begin{pmatrix} 9 \\ -3 \end{pmatrix}$ to give triangle B.

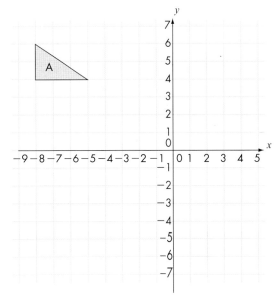

       Triangle B is then enlarged by a scale factor –2 about the origin to give triangle C.

       Draw triangles B and C on the diagram.

     b Describe fully the single transformation that maps triangle C onto triangle A.

7.5 Bisectors

Homework 7H

1   Draw a line 8 cm long and bisect it. Check your accuracy by measuring each half.

2   **a** Draw any triangle.

     b On each side construct the line bisector. Your line bisectors should intersect at the same point.

     c Using this point as the centre, draw a circle that passes through the three vertices of the triangle.

3   **a** Draw a circle with a radius of about 4 cm.

     b Draw a quadrilateral inside the circle so that the vertices of the quadrilateral touch its circumference.

     c Bisect two of the sides of the quadrilateral. Your bisectors should meet at the centre of the circle.

4   **a** Draw any angle.

     b Construct the angle bisector.

     c Check your accuracy by measuring each half.

5   The diagram shows a park with two ice-cream sellers A and B. People always go to the ice-cream seller nearest to them. Shade the region of the park from which people go to ice-cream seller B.

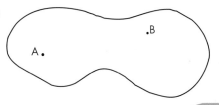

6 Using only a pencil, a straight edge and a pair of compasses, construct:

    a an angle of 15 degrees

    b an angle of 75 degrees.

7 If I construct all the angle bisectors of any triangle, they will meet at a point.

    Show that if I draw a circle with this as the centre, the circle will just touch each side of the triangle.

7.6 Defining a locus

Homework 7I

1 A is a fixed point. Sketch the locus of the point P when AP > 3 cm and AP < 6 cm.

2 A and B are two fixed points 4 cm apart. Sketch the locus of the point P for these situations:

    a AP < BP   **b** P is always within 3 cm of A and within 2 cm of B.

3 A fly is tethered by a length of spider's web that is 1 m long. Describe the locus of the fly's movement.

4 ABC is an equilateral triangle of side 4 cm. In each of the following loci, the point P moves only inside the triangle. Sketch the locus in each case.

    a AP = BP   **b** AP < BP

    c CP < 2 cm   **d** CP > 3 cm and BP > 3 cm

5 A wheel rolls around the inside of a square. Sketch the locus of the centre of the wheel.

6 The same wheel rolls around the outside of the square. Sketch the locus of the centre of the wheel.

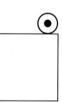

7 On a piece of plain paper, mark three points A, B and C, about 5 to 7 cm away from each other.

    Find the locus of point P when:

    a P is always closer to a point A than a point B

    b P is always the same distance from points B and C.

8 Sketch the locus of a point on the rim of a bicycle wheel as it makes three revolutions along a flat road.

7.7 Loci problems

Homework 7J

For Questions 1 to 3, you should start by sketching the picture given in each question before drawing the locus accurately. The scale for each question is given.

1 A goat is tethered by a rope, 10 m long, to a stake that is 2 m from each side of a field. What is the locus of the area that the goat can graze? Use a scale of 1 cm to 2 m.

2 A cow is tethered to a rail at the top of a fence 4 m long. The rope is 4 m long. Sketch the area that the cow can graze. Use a scale of 1 cm to 2 m.

3 A horse is tethered to a corner of a shed, 3 m by 1 m. The rope is 4 m long. Sketch the area that the horse can graze. Use a scale of 1 cm to 1 m.

4 Two ships, A and B, which are 7 km apart, both hear a distress signal from a fishing boat. The fishing boat is less than 4 km from ship A and less than 4.5 km from ship B. A helicopter pilot sees that the fishing boat is nearer to ship A than to ship B. Use accurate construction to show the region which contains the fishing boat. Shade this region.

5 The locus of a point is described as:

5 cm away from point A

equidistant from both points A and B.

Which of the following could be true?

a The locus is an arc.

b The locus is just two points.

c The locus is a straight line.

d The locus is none of these.

For Questions 6 to 9, you should use a copy of the map on this page. For each question, trace the map and mark on those points that are relevant to that question.

7 Geometry and measures: Transformations, constructions and loci

6 A radio station broadcasts from Birmingham with a range that is just far enough to reach York. Another radio station broadcasts from Glasgow with a range that is just far enough to reach Newcastle.

    a Sketch the area to which each station can broadcast.

    b Will the Birmingham station broadcast as far as Norwich?

    c Will the two stations interfere with each other?

7 An air traffic control centre is to be built in Newcastle. If it has a range of 200 km, will it cover all the area of Britain north of Sheffield and south of Glasgow?

8 There are plans to build a new radio transmitter so that it is the same distance from Exeter, Norwich and Newcastle.

    a Draw the perpendicular bisectors of the lines joining these three places and hence find its proposed location.

    b The radio transmitter will cause problems if it is built within 50 km of Birmingham. Will the proposed location cause problems?

9 Three radio stations pick up a distress call from a boat in the North Sea.

The station at Norwich can tell from the strength of the signal that the boat is within 150 km of the station. The station at Sheffield can tell that the boat is between 100 and 150 km from Sheffield.

If these two reports are correct, state the least and greatest possible distance of the boat from the helicopter station at Newcastle?

7.8 Plans and elevations

Homework 7K

1 Draw the plan, front elevation and side elevation for each shape.

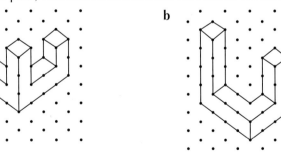

    a   **b**

2 This 3D shape is made from cubes.

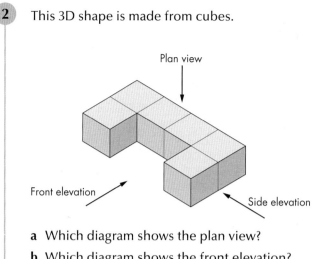

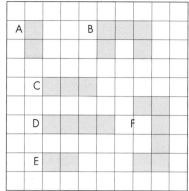

a Which diagram shows the plan view?

b Which diagram shows the front elevation?

c Which diagram shows the side elevation?

3 On an isometric grid, draw the 3D shape shown by this plan, front elevation and side elevation.

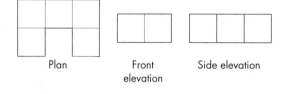

4 Draw an accurate plan, front elevation and side elevation for a 7 cm-long regular hexagonal prism with side length 3 cm.

5 The diagram shows the front elevation of a storage tank with a uniform cross-section. The storage tank is 4 metres deep.

Draw the plan and side elevation of the tank using a scale of 2 cm to 1 m.

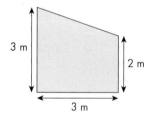

6 Draw an accurate plan, front elevation and side elevation these shapes.

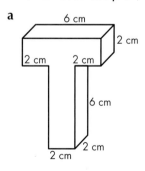

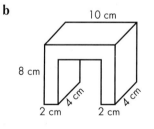

8 Algebra: Algebraic manipulation

8.1 Basic algebra

Homework 8A

1 Find the value of $4x + 3$ when: **a** $x = 3$ **b** $x = 6$ **c** $x = 11$.

2 Find the value of $3k - 1$ when: **a** $k = 2$ **b** $k = 5$ **c** $k = 10$.

3 Find the value of $4 + t$ when: **a** $t = 5$ **b** $t = 8$ **c** $t = 15$.

4 Evaluate $14 - 3f$ when: **a** $f = 4$ **b** $f = 6$ **c** $f = 10$.

5 Find the value of $\dfrac{4d - 7}{2}$ when: **a** $d = 2$ **b** $d = 5$ **c** $d = 15$.

6 Find the value of $5x + 2$ when: **a** $x = -2$ **b** $x = -1$ **c** $x = 21.5$.

7 Evaluate $4w - 3$ when: **a** $w = -2$ **b** $w = -3$ **c** $w = 2.5$.

8 Evaluate $10 - x$ when: **a** $x = -3$ **b** $x = -6$ **c** $x = 4.6$.

9 Find the value of $5t - 1$ when: **a** $t = 2.4$ **b** $t = -2.6$ **c** $t = 0.05$.

10 Evaluate $11 - 3t$ when: **a** $t = 2.5$ **b** $t = -2.8$ **c** $t = 0.99$.

11 Where $H = a^2 + c^2$, find the value of H when:
a $a = 3$ and $c = 4$ **b** $a = 5$ and $c = 12$.

12 Where $K = m^2 - n^2$, find the value of K when:
a $m = 5$ and $n = 3$ **b** $m = -5$ and $n = -2$.

13 Where $P = 100 - n^2$, find the value of P when:
a $n = 7$ **b** $n = 8$ **c** $n = 9$.

14 Where $D = 5x - y$, find the value of D when:
a $x = 4$ and $y = 3$ **b** $x = 5$ and $y = -3$.

15 Where $T = y(2x + 3y)$, find the value of T when:
a $x = 8$ and $y = 12$ **b** $x = 5$ and $y = 7$.

16 Where $m = w(t^2 + w^2)$, find the value of m when:
a $t = 5$ and $w = 3$ **b** $t = 8$ and $w = 7$.

17 Two of the first recorded units of measurement were the *cubit* and the *palm*.

The cubit is the distance from the fingertip to the elbow and the palm is the distance across the hand.

A cubit is four and a half palms.

The actual length of a cubit varied throughout history, but it is now accepted to be 54 cm.

Noah's Ark is recorded as being 300 cubits long by 50 cubits wide by 30 cubits high.

What are the dimensions of the Ark in metres?

18 In this algebraic magic square, every row, column and diagonal should add up and simplify to $9a + 6b + 3c$.

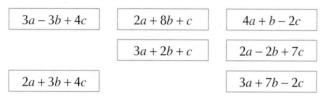

$3a - 3b + 4c$	$2a + 8b + c$	$4a + b - 2c$
	$3a + 2b + c$	$2a - 2b + 7c$
$2a + 3b + 4c$		$3a + 7b - 2c$

a Copy and complete the magic square.

b Calculate the value of the 'magic number' when $a = 2$, $b = 3$ and $c = 4$.

19 The rule for converting degrees Fahrenheit (F) into degrees Celsius (C) is:
$C = \dfrac{5}{9}(F - 32)$.

a Use this rule to convert 68°F into degrees Celsius.

b Which of the following is the rule for converting degrees Celsius into degrees Fahrenheit?

$$F = \frac{9}{5}(C + 32) \qquad F = \frac{5}{9}C + 32 \qquad F = \frac{9}{5}C + 32 \qquad F = \frac{9}{5}C - 32$$

20 The formula for the cost of water used by a household each quarter is:

£32.40 + £0.003 per litre of water used.

A family uses 450 litres of water each day.

a How much is their total bill per quarter? (Take a quarter to be 91 days.)

b The family pay a direct debit of £45 per month towards their water costs.

By how much will they be in credit or debit after the quarter?

21 Work out the value of each expression when $x = 17.4$, $y = 28.2$ and $z = 0.6$.

a $x + \dfrac{y}{z}$ **b** $\dfrac{x + y}{z}$ **c** $\dfrac{x}{z} + y$

22 a Laser printer cartridges cost £75 and print approximately 2500 pages.

What is the approximate cost of ink per page printed?

b A printing specialist uses a laser printer of this type. He charges a fixed rate of £4.50 to set up the design and five pence for every page.

Show that his profit on a print run of x pages is, in pounds, $4.5 + 0.02x$.

c How much profit will the printing specialist make if he prints 2000 race entry forms for a running club?

23 State whether each of the following is an expression, equation, formula or identity.

a $5x^2 - 2x + 1$ **b** $V = IR$

c $3(x + 2) = 3x + 6$ **d** $3x - 2 = 0$

Homework 8B

1 Expand these expressions.

a $3(4 + m)$	**b** $6(3 + p)$	**c** $4(4 - y)$	**d** $3(6 + 7k)$
e $4(3 - 5f)$	**f** $2(4 - 23w)$	**g** $7(g + h)$	**h** $4(2k + 4m)$
i $6(2d - n)$	**j** $t(t + 5)$	**k** $m(m + 4)$	**l** $k(k - 2)$
m $g(4g + 1)$	**n** $y(3y - 21)$	**o** $p(7 - 8p)$	**p** $2m(m + 5)$
q $3t(t - 2)$	**r** $3k(5 - k)$	**s** $2g(4g + 3)$	**t** $4h(2h - 3)$
u $2t(6 - 5t)$	**v** $4d(3d + 5e)$	**w** $3y(4y + 5k)$	**x** $6m^2(3m - p)$
y $y(y^2 + 7)$	**z** $h(h^3 + 9)$	**aa** $k(k^2 - 4)$	**bb** $3t(t^2 + 3)$
cc $5h(h^3 - 2)$	**dd** $4g(g^3 - 3)$	**ee** $5m(2m^2 + m)$	**ff** $2d(4d^2 - d^3)$
gg $4w(3w^2 + t)$	**hh** $3a(5a^2 - b)$	**ii** $2p(7p^3 - 8m)$	**jj** $m^2(3 + 5m)$
kk $t^3(t + 3t)$	**ll** $g^2(4t - 3g^2)$	**mm** $2t^2(7t + m)$	**nn** $3h^2(4h + 5g)$

2 An approximate rule for converting degrees Fahrenheit into degrees Celsius is:

$C = 0.5(F - 30)$.

a Use this rule to convert 22°F into degrees Celsius.

b Which of the following is an approximate rule for converting degrees Celsius into degrees Fahrenheit?

$F = 2(C + 30)$ $F = 0.5(C + 30)$ $F = 2(C + 15)$ $F = 2(C - 15)$

3 Match the equivalent algebraic expressions. One has been done for you.

$2y$

$y + y$ $5(y - 2)$

$3y + 6$ $y \times y$

y^2 $5y - 2$

$5y - 10$ $3y + 2$

$3(y + 2)$

4 The expansion $3(4x + 8y) = 12x + 24y$.

Write down two other expansions that give an answer of $12x + 24y$.

5 Simplify these expressions.

a $5t + 4t$	**b** $4m + 3m$	**c** $6y + y$	**d** $2d + 3d + 5d$
e $7e - 5e$	**f** $6g - 3g$	**g** $3p - p$	**h** $5t - t$
i $t^2 + 4t^2$	**j** $5y^2 - 2y^2$	**k** $4ab + 3ab$	**l** $5a^2d - 4a^2d$

6 Expand and simplify.

a $3(2 + t) + 4(3 + t)$ **b** $6(2 + 3k) + 2(5 + 3k)$ **c** $5(2 + 4m) + 3(1 + 4m)$

d $3(4 + y) + 5(1 + 2y)$ **e** $5(2 + 3f) + 3(6 - f)$ **f** $7(2 + 5g) + 2(3 - g)$

g $4(3 + 2h) - 2(5 + 3h)$ **h** $5(3g + 4) - 3(2g + 5)$ **i** $3(4y + 5) - 2(3y + 2)$

j $3(5t + 2) - 2(4t + 5)$ **k** $5(5k + 2) - 2(4k - 3)$ **l** $4(4e + 3) - 2(5e - 4)$

m $m(5 + p) + p(2 + m)$ **n** $k(4 + h) + h(5 + 2k)$ **o** $t(1 + 2n) + n(3 + 5t)$

p $p(5q + 1) + q(3p + 5)$ **q** $2h(3 + 4j) + 3j(h + 4)$ **r** $3y(4t + 5) + 2t(1 + 4y)$

s $t(2t + 5) + 2t(4 + t)$ **t** $3y(4 + 3y) + y(6y - 5)$ **u** $5w(3w + 2) + 4w(3 - w)$

v $4p(2p + 3) - 3p(2 - 3p)$ **w** $4m(m - 1) + 3m(4 - m)$ **x** $5d(3 - d) + d(2d - 1)$

y $5a(3b + 2a) + a(2a^2 + 3c)$ **z** $4y(3w + y^2) + y(3y - 4t)$

7 Adult tickets for a concert cost £x and children's tickets cost £y.

At the afternoon show there were 40 adults and 160 children.

At the evening show there were 60 adults and 140 children.

a Write down an expression for the total amount of money taken on that day in terms of x and y.

b The daily expense for putting on the show is £2200. If $x = 12$ and $y = 9$, how much profit did the theatre make that day?

8 Don wrote the following.

$2(3x - 1) + 5(2x + 3) = 5x - 2 + 10x + 15 = 15x - 13$

Don has made two mistakes in his working. Explain the mistakes that Don has made.

9 An internet site sells CDs. They cost £$(x + 0.75)$ each for the first five and then £$(x + 0.25)$ for any orders over five.

a Zoe buys eight CDs. Which of the following expressions represents how much Zoe will pay?

     i $8(x + 0.75)$ **ii** $5(x + 0.75) + 3(x + 0.25)$

     iii $3(x + 0.75) + 5(x + 0.25)$ **iv** $8(x + 0.25)$

b If $x = 5$, how much will Zoe pay?

8.2 Factorisation

Homework 8C

1 Factorise these expressions.

a $9m + 12t$ **b** $9t + 6p$ **c** $4m + 12k$ **d** $4r + 6t$

e $2mn + 3m$ **f** $4g^2 + 3g$ **g** $4w - 8t$ **h** $10p - 6k$

i $12h - 10k$ **j** $4mp + 2mk$ **k** $4bc + 6bk$ **l** $8ab + 4ac$

m $3y^2 + 4y$ **n** $5t^2 - 3t$ **o** $3d^2 - 2d$ **p** $6m^2 - 3mp$

q $3p^2 + 9pt$ **r** $8pt + 12mp$ **s** $8ab - 6bc$ **t** $4a^2 - 8ab$

u $8mt - 6pt$ **v** $20at^2 + 12at$ **w** $4b^2c - 10bc$ **x** $4abc + 6bed$

y $6a^2 + 4a + 10$ **z** $12ab + 6bc + 9bd$ **aa** $6t^2 + 3t + at$

bb $96mt^2 - 3mt + 69m^2t$ **cc** $6ab^2 + 2ab - 4a^2b$ **dd** $5pt^2 + 15pt + 5p^2t$

2 Factorise these expressions where possible. List those that do not factorise.

a $5m - 6t$ **b** $3m + 2mp$ **c** $t^2 - 5t$ **d** $6pt + 5ab$

e $8m^2 - 6mp$ **f** $a^2 + c$ **g** $3a^2 - 7ab$ **h** $4ab + 5cd$

i $7ab - 4b^2c$ **j** $3p^2 - 4t^2$ **k** $6m^2t + 9t^2m$ **l** $5mt + 3pn$

3 An ink cartridge is priced at £9.99.

The shop has a special offer of 20% off if you buy five or more. 20% of £9.99 is £1.99.

Tom wants six cartridges. Tess wants eight cartridges.

Tom writes down the calculation $6 \times 9.99 - 6 \times 1.99$ to work out how much he must pay.

Tess writes down the calculation $8 \times (9.99 - 1.99)$ to work out how much she must pay.

Both calculations are correct.

a Who has the easier calculation and why?

b How much will each of them pay for their cartridges?

4 **a** Factorise these expressions.

    i $4x + 3 + 5x - 7 - 8x$ **ii** $3x - 12$ **iii** $x^2 - 4x$

b What do all the answers in **a** have in common?

5 A class of students were asked to add up all the numbers from 1 to 100 (i.e. $1 + 2 + 3 + 4 + \ldots + 98 + 99 + 100$).

Two minutes later, a student said she had the correct answer.

The teacher asked the student to show the class her method.

The student wrote:

$(1 + 100) + (2 + 99) + (3 + 98) + \ldots (50 + 51) = 50 \times 101$

a Show that this gives the correct answer.

b What is the sum of all the numbers from 1 to 100?

8.3 Quadratic expansion

Homework 8D

1 Use the expansion method to expand these expressions.

a $(x + 2)(x + 5)$ **b** $(t + 3)(t + 2)$ **c** $(w + 4)(w + 1)$ **d** $(m + 6)(m + 2)$

e $(k + 2)(k + 4)$ **f** $(a + 3)(a + 1)$ **g** $(x + 3)(x - 1)$ **h** $(t + 6)(t - 4)$

i $(w + 2)(w - 3)$ **j** $(f + 1)(f - 4)$ **k** $(g + 2)(g - 5)$ **l** $(y + 5)(y - 2)$

m $(x - 4)(x + 3)$ **n** $(p - 3)(p + 2)$ **o** $(k - 5)(k + 1)$

1 Use the FOIL method to expand these expressions.

   a $(y - 3)(y + 6)$ **b** $(a - 2)(a + 4)$ **c** $(t - 4)(t + 5)$ **d** $(x - 3)(x - 2)$

   e $(r - 4)(r - 1)$ **f** $(m - 1)(m - 7)$ **g** $(g - 5)(g - 3)$ **h** $(h - 6)(h - 2)$

   i $(n - 2)(n - 8)$ **j** $(4 + x)(3 + x)$ **k** $(5 + t)(4 - t)$ **l** $(2 - b)(6 + b)$

   m $(7 - y)(5 - y)$ **n** $(3 + p)(p - 2)$ **o** $(3 - k)(k - 5)$

Homework 8F

1 The expansions in this question follow a pattern. Use the box method to work out the first few and try to spot the pattern that will allow you to immediately write down the answers to the rest.

   a $(x + 1)(x - 1)$ **b** $(t + 2)(t - 2)$ **c** $(y + 3)(y - 3)$ **d** $(k - 3)(k + 3)$

   e $(h - 1)(h + 1)$ **f** $(3 + x)(3 - x)$ **g** $(7 + t)(7 - t)$ **h** $(4 - y)(4 + y)$

   i $(a + b)(a - b)$ **j** $(6 + x)(6 - x)$

2 This rectangle is made up of four parts with areas of x^2, $5x$, $4x$ and 20 square units.

   Work out expressions in terms of x for the sides of the rectangle.

x^2	$5x$
$4x$	20

3 This square has an area of x^2 square units.

   It is split into four rectangles.

   a Copy and complete the table to show the dimensions and area of each rectangle.

Rectangle	Length	Height	Area
A	$x - 2$	$x - 3$	$(x - 2)(x - 3)$
B			
C			
D			

   b Add together the areas of rectangles B, C and D.

   Expand any brackets and collect together like terms.

   c Use the results to show that why $(x - 2)(x - 3) = x^2 - 5x + 6$.

4 **a** Expand $(x - 2)(x + 2)$.

   b Use the result in **a** to write down the answers to these. (Do not use a calculator or long multiplication.)

   i 98×102 **ii** 198×202

1 Expand these expressions.

a $(3x + 4)(4x + 2)$ **b** $(2y + 1)(3y + 2)$ **c** $(4t + 2)(3t + 6)$

d $(3t + 2)(2t - 1)$ **e** $(6m + 1)(3m - 2)$ **f** $(5k + 3)(4k - 3)$

g $(4p - 5)(3p + 4)$ **h** $(6w + 1)(3w + 4)$ **i** $(3a - 4)(5a + 1)$

j $(5r - 2)(3r - 1)$ **k** $(4g - 1)(3g - 2)$ **l** $(3d - 2)(4d + 1)$

m $(3 + 4p)(5 + 4p)$ **n** $(3 + 2t)(5 + 3t)$ **o** $(2 + 5p)(3p + 1)$

p $(7 + 4t)(3 - 2t)$ **q** $(5 + 2n)(4 - n)$ **r** $(3 + 4f)(5f - 1)$

s $(2 - 3q)(5 + 4q)$ **t** $(3 - p)(2 + 3p)$ **u** $(5 - 3t)(4t + 1)$

v $(5 - 4r)(3 - 4r)$ **w** $(4 - x)(1 - 5x)$ **x** $(2 - 7m)(2m - 3)$

y $(x + y)(3x + 5y)$ **z** $(4y + t)(3y - 4t)$ **aa** $(5x - 3y)(5x + y)$

bb $(x - 2y)(x - 3y)$ **cc** $(4m - 3p)(m + 5p)$ **dd** $(t - 4k)(3t - k)$

2 **a** Expand $(x + 1)(x + 1)$

b Expand $(x - 1)(x - 1)$

c Expand $(x + 1)(x - 1)$

d Use the results in parts **a**, **b** and **c** to show that $(p - q)^2 \equiv p^2 - 2pq + q^2$ is an identity.

> **Hints and tips** Take $p = x + 1$ and $q = x - 1$.

3 Imagine a square of side $2a$ units with a square of side x units cut from one corner.

a What is the area remaining after the small square is cut away?

b The remaining area is cut into three rectangles, A, B and C, and rearranged as shown.

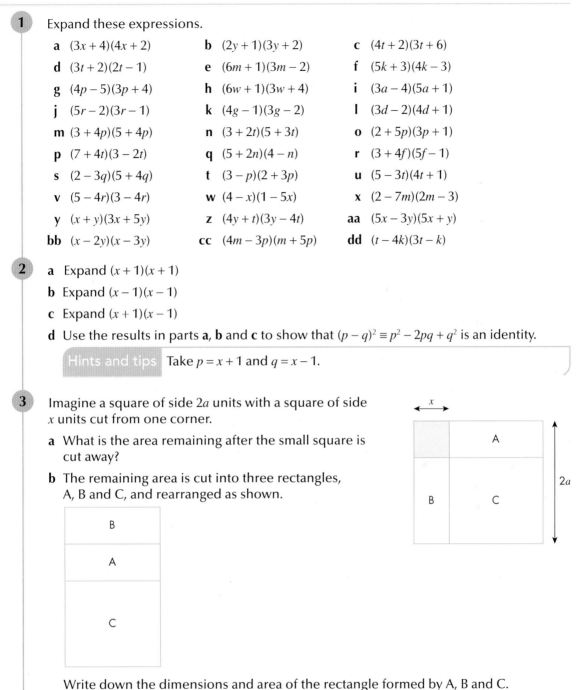

Write down the dimensions and area of the rectangle formed by A, B and C.

c Show that $4a^2 - bx^2 = (2a + x)(2a - x)$.

4 **a** Without expanding the brackets, match each expression on the left with an expression on the right. One is done for you.

$(4x - 3)(3x + 2)$ $12x^2 - x - 1$

$(2x - 1)(6x - 1)$ $12x^2 - x - 6$

$(6x - 1)(2x + 3)$ $12x^2 + 37x + 3$

$(4x + 1)(3x - 1)$ $12x^2 - 8x + 1$

$(x + 3)(12x + 1)$ $12x^2 + 16x - 3$

   b Taking any expression on the left, show how you can match it with an expression on the right without expanding the brackets.

8.4 Expanding squares

Homework 8H

1 Expand these squares and simplify.

a $(x + 4)^2$ **b** $(m + 3)^2$ **c** $(5 + t)^2$ **d** $(2 + p)^2$

e $(m - 2)^2$ **f** $(t - 4)^2$ **g** $(3 - m)^2$ **h** $(6 - k)^2$

i $(2x + 1)^2$ **j** $(3t + 2)^2$ **k** $(1 + 4y)^2$ **l** $(2 + m)^2$

m $(3t - 2)^2$ **n** $(2x - 1)^2$ **o** $(1 - 4t)^2$ **p** $(5 - 4r)^2$

q $(a + b)^2$ **r** $(x - y)^2$ **s** $(3t + y)^2$ **t** $(m - 2n)^2$

u $(x + 3)^2 - 4$ **v** $(x - 4)^2 - 25$ **w** $(x + 5)^2 - 36$ **x** $(x - 1)^2 - 1$

2 A teacher asks her class to expand $(4x - 1)^2$.

Bernice's answer is $16x^2 - 1$.

Piotr's answer is $16x^2 + 8x - 1$.

   a Explain the mistakes that Bernice has made.

   b Explain the mistakes that Piotr has made.

   c Work out the correct answer.

8.5 More than two binomials

Homework 8I

1 Expand these expressions and then simplify.

a $(x + 2)(x + 3)(x + 4)$ **b** $(x + 1)(x - 3)(x + 2)$

c $(x - 3)(x + 3)(x - 5)$ **d** $(x - 4)^2(x + 3)$

e $(x^2 + 2x + 1)(x - 3)$

2 Expand these expressions and then simplify.

a $(x + 1)^3$ **b** $(x - 2)^3$ **c** $(x + 4)^3$

3 **a** Expand $(x + a)(x + b)(x + c)$.

   b If $(x + 1)(x - 4)(x + 3) = x^3 + px^2 + qx + r$, use your answer from a to find the values of p, q and r.

4 A cuboid has edges of $(x - 1)$ cm, $(x + 4)$ cm and $(x - 9)$ cm.

Find simplified expressions for:

a the volume of the cuboid

b the surface area of the cuboid.

5 **a** Expand $(x + 2)^3$.

b Find the value of 2.01^3 without using a calculator.

6 Expand these expressions and then simplify.

a $(2x - 3)(x + 4)(x - 1)$ **b** $(x + 2)^2(3x - 1)$

7 Expand these expressions and then simplify.

a $(2x - 4)(3x + 1)(4x - 2)$ **b** $(2x - 4)^2(2x - 8)$

8.6 Quadratic factorisation

Homework 8J

1 Factorise these expressions.

a $x^2 + 7x + 6$	**b** $t^2 + 4t + 4$	**c** $m^2 + 11m + 10$	**d** $k^2 + 11k + 24$
e $p^2 + 10p + 24$	**f** $r^2 + 11r + 18$	**g** $w^2 + 9w + 18$	**h** $x^2 + 8x + 12$
i $a^2 + 13a + 12$	**j** $k^2 - 10k + 21$	**k** $f^2 - 22f + 21$	**l** $b^2 + 35b + 96$
m $t^2 + 5t + 6$	**n** $m^2 - 5m + 4$	**o** $p^2 - 7p + 10$	**p** $x^2 - 13x + 36$
q $c^2 - 12c + 32$	**r** $t^2 - 15t + 36$	**s** $y^2 - 14y + 48$	**t** $j^2 - 19j + 48$
u $p^2 + 8p + 15$	**v** $y^2 + y - 6$	**w** $t^2 + 7t - 8$	**x** $x^2 + 9x - 10$
y $m^2 - m - 12$	**z** $r^2 + 6r - 7$	**aa** $n^2 - 7n - 18$	**bb** $m^2 - 20m - 44$
cc $w^2 - 5w - 24$	**dd** $t^2 + t - 90$	**ee** $x^2 - x - 72$	**ff** $t^2 - 18t - 63$
gg $d^2 - 2d + 1$	**hh** $y^2 + 29y + 100$	**ii** $t^2 - 10t + 16$	**jj** $m^2 - 30m + 81$
kk $x^2 - 30x + 144$	**ll** $d^2 - 4d - 12$	**mm** $t^2 + t - 20$	**nn** $q^2 + q - 56$
oo $p^2 - p - 2$	**pp** $v^2 - 2v - 35$	**qq** $t^2 - 4t + 3$	**rr** $m^2 + 3m - 4$

2 This rectangle is made up of four parts. Two of the parts have areas of x^2 and 9 square units.

The sides of the rectangle are of the form $x + a$ and $x + b$.

There are two possible answers for a and b.

Work out both answers and copy and complete the areas in the other parts of the rectangle.

3 **a** Expand $(x - a)(x - b)$.

b If $x^2 - 9x + 18 = (x - p)(x - q)$, use your answer to part **a** to write down the values of:

     i $p + q$ **ii** pq.

c Show how you can tell that $x^2 - 18x + 9$ will not factorise.

Each of the expressions in questions 1 to 12 is the difference of two squares. Factorise them.

1
a $x^2 - 81$ b $t^2 - 36$ c $4 - x^2$
d $81 - t^2$ e $k^2 - 400$ f $64 - y^2$

2
a $x^2 - y^2$ b $a^2 - 9b^2$ c $9x^2 - 25y^2$
d $9x^2 - 16$ e $100t^2 - 4w^2$ f $36a^2 - 49b^2$

g Factorise the denominator and hence simplify $\dfrac{2a - 3}{4a^2 - 9}$.

3
a A square has a side of $2x$ units.

What is the area of the square?

b A rectangle, A, 3 units wide, is cut from the square and placed at the side of the remaining rectangle, B.

A square, C, is then cut from the bottom of rectangle A to leave a final rectangle, D.

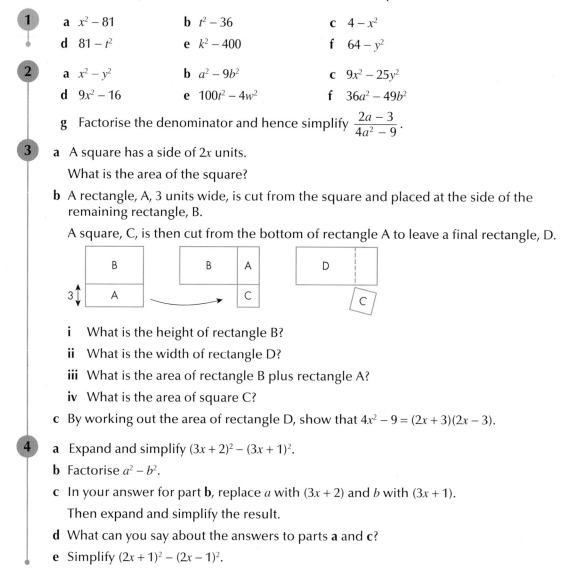

i   What is the height of rectangle B?

ii What is the width of rectangle D?

iii What is the area of rectangle B plus rectangle A?

iv What is the area of square C?

c By working out the area of rectangle D, show that $4x^2 - 9 = (2x + 3)(2x - 3)$.

4
a Expand and simplify $(3x + 2)^2 - (3x + 1)^2$.

b Factorise $a^2 - b^2$.

c In your answer for part **b**, replace a with $(3x + 2)$ and b with $(3x + 1)$.

Then expand and simplify the result.

d What can you say about the answers to parts **a** and **c**?

e Simplify $(2x + 1)^2 - (2x - 1)^2$.

8.7 Factorising $ax^2 + bx + c$

1 Factorise these expressions.

a $3x^2 + 4x + 1$ b $3x^2 - 2x - 1$ c $4x^2 + 8x + 3$ d $2x^2 + 7x + 3$
e $15x^2 + 13x + 2$ f $4x^2 + 4x - 3$ g $6x^2 - 7x + 2$ h $8x^2 - 8x - 6$
i $8x^2 - 13x - 6$ j $6x^2 - 13x + 2$ k $6x^2 - 2x$ l $6x^2 + 11x - 2$

2 This rectangle is made up of four parts, with areas of $6x^2$, $3x$, $8x$ and 4 square units.

$6x^2$	$3x$
$8x$	4

Work out expressions in terms of x for the sides of the rectangle.

3 Three students are asked to factorise the expression $4x^2 + 4x - 8$.

These are their answers:

Adriana   Ben   Cara

$(2x + 4)(2x - 2)$ $(4x + 8)(x - 1)$ $(x + 2)(4x - 4)$

All the answers are correctly factorised.

a Show that one quadratic expression can have three different factorisations.

b Which of the following is the most complete factorisation? Give reasons for your choice.

$2(x + 2)(2x - 2)$ $4(x + 2)(x - 1)$ $2(2x + 4)(x - 1)$

8.8 Changing the subject of a formula

Homework 8M

1 A restaurant has a large oven that can cook up to 10 chickens at a time.

The chef uses the formula:

$T = 10n + 55$

to calculate the length of time (T) it takes to cook n chickens

A large group is booked for a chicken dinner at 7 pm. They will need a total of eight chickens.

a It takes 15 minutes to get the chickens out of the oven and prepare them for serving.

At what time should the chef put the eight chickens into the oven?

b Rearrange the formula to make n the subject

c Another large group is booked for 8 pm the following day. The chef calculates she will need to put the chickens in the oven at 5.50 pm.

How many chickens is the chef cooking for this party?

2 Fern notices that the price of six coffees is 90 pence less than the price of nine teas.

Let the price of a coffee be x pence and the price of a tea be y pence.

a Express the cost of a tea, y, in terms of the price of a coffee, x.

b If the price of a coffee is £1.20, how much is a tea?

3 Distance, speed and time are connected by the formula:

distance = speed × time.

A delivery driver drove 90 miles at an average speed of 60 miles per hour.

On the return journey, he was held up at some road works for 30 minutes.

What was his average speed on the return journey?

4 $y = mx + c$ **a** Make c the subject.   **b** Express x in terms of y, m and c.

5 $v = u - 10t$ **a** Make u the subject.   **b** Express t in terms of v and u.

6 $T = 2x + 3y$ **a** Express x in terms of T and y.   **b** Make y the subject.

7 $p = q^2$ Make q the subject.

8 $p = q^2 - 3$ Make q the subject.

9 $a = b^2 + c$ Make b the subject.

10 A rocket is fired vertically upwards with an initial velocity of u metres per second. After t seconds the rocket's velocity, v metres per second, is given by the formula $v = u + gt$, where g is a constant.

a Calculate v when $u = 120$, $g = -9.8$ and $t = 6$.

b Rearrange the formula to express t in terms of v, u, and g.

c Calculate t when $u = 100$, $g = -9.8$ and $v = 17.8$.

9 Geometry and measures: Length, area and volume

9.1 Circumference and area of a circle

Homework 9A

1. Find the circumference of each circle. Give your answers to 1 dp.
   a Diameter 3 cm
   b Radius 5 cm
   c Radius 8 m
   d Diameter 14 cm
   e Diameter 6.4 cm
   f Radius 3.5 cm

2. John runs twice round a circular track with a radius of 50 m. How far has he run? Give your answers in terms of π.

3. A rolling pin has a diameter of 5 cm.
   a What is the circumference of the rolling pin?
   b How many complete revolutions does it make when rolling a length of 30 cm?

4. A bicycle has wheels with a radius of 28 cm. How many complete revolutions will each wheel make in a journey of 3 km?

5. Calculate the area of each circle. Give your answers to parts **a** to **d** to 1 decimal place and to parts **e** and **f** in terms of π.
   a Diameter 14 cm
   b Radius 9 cm
   c Radius 21 cm
   d Diameter 0.9 cm
   e Radius 4 cm
   f Diameter 2 m

6. What is the total perimeter of a semicircle of diameter 7 cm? Give your answer to 1 dp.

7. What is the total perimeter of a semicircle of radius 6 cm? Give your answer in terms of π.

8. A circle has a circumference of 12 cm. What is its diameter?

9. A garden has a circular lawn of diameter 20 m. There is a path 1 m wide all the way round the circumference. What is the area of this path?

10. Calculate the area of a semicircle with a diameter of 15 cm. Give your answer to 1 dp.

11. A circle has an area of 50 m². What is its radius?

12. I draw a circle with a circumference of 25 cm. What is the area of this circle?

13. Calculate the area of this shape.

25 cm

20 cm

14 Jane walked around a circular lawn. She took 153 paces. Each of her paces was about 42 cm. What is the area of the lawn?

15 The wheels of a bicycle have a diameter of 70 cm. How many metres will the bicycle travel if each wheel makes 50 revolutions?

9.2 Area of a parallelogram

Homework 9B

1 Calculate the area of each parallelogram.

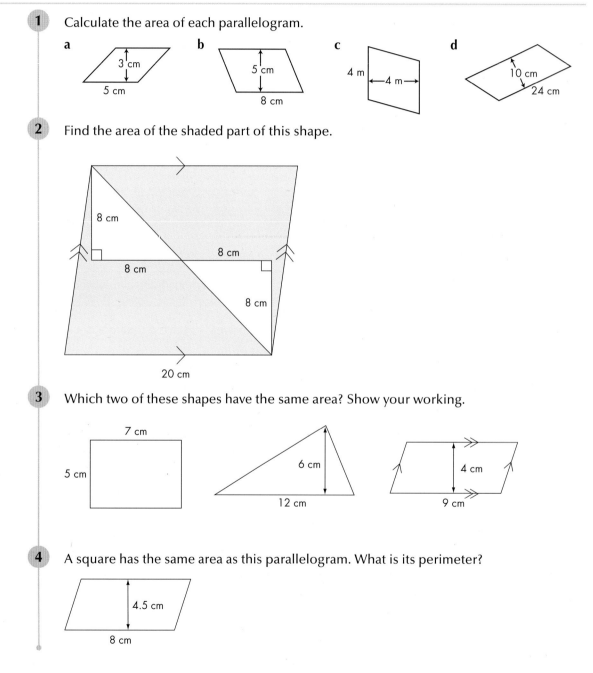

a

3 cm
5 cm

b

5 cm
8 cm

c

4 m
4 m

d

10 cm
24 cm

2 Find the area of the shaded part of this shape.

8 cm

8 cm

8 cm

8 cm

20 cm

3 Which two of these shapes have the same area? Show your working.

7 cm
5 cm

6 cm
12 cm

4 cm
9 cm

4 A square has the same area as this parallelogram. What is its perimeter?

4.5 cm
8 cm

9.3 Area of a trapezium

Homework 9C

1 Calculate the perimeter and the area of each trapezium.

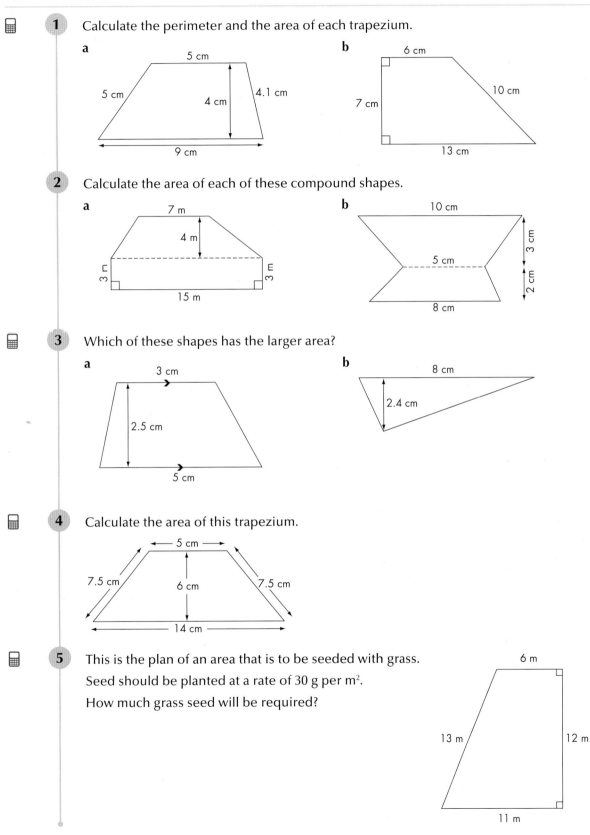

a

5 cm

5 cm

4 cm

4.1 cm

9 cm

b

6 cm

7 cm

10 cm

13 cm

2 Calculate the area of each of these compound shapes.

a

7 m

4 m

3 m

3 m

15 m

b

10 cm

5 cm

3 cm

2 cm

8 cm

3 Which of these shapes has the larger area?

a

3 cm

2.5 cm

5 cm

b

8 cm

2.4 cm

4 Calculate the area of this trapezium.

5 cm

7.5 cm

6 cm

7.5 cm

14 cm

5 This is the plan of an area that is to be seeded with grass.
Seed should be planted at a rate of 30 g per m².
How much grass seed will be required?

6 m

13 m

12 m

11 m

6 A trapezium has an area of 100 cm². The parallel sides are 17 cm and 23 cm in length. How far apart are the parallel sides?

7 Calculate the area of the shaded part in each diagram.

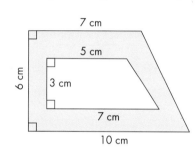

a
6 cm
4 cm
5 cm
8 cm
2 cm
9 cm

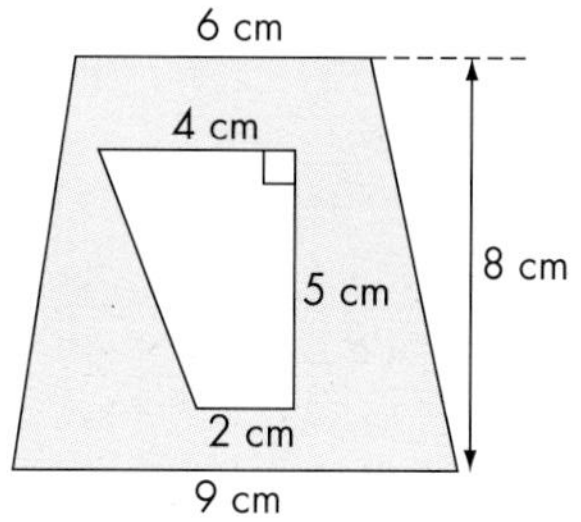

b
4 cm
3 cm
2 cm
6 cm
5 cm

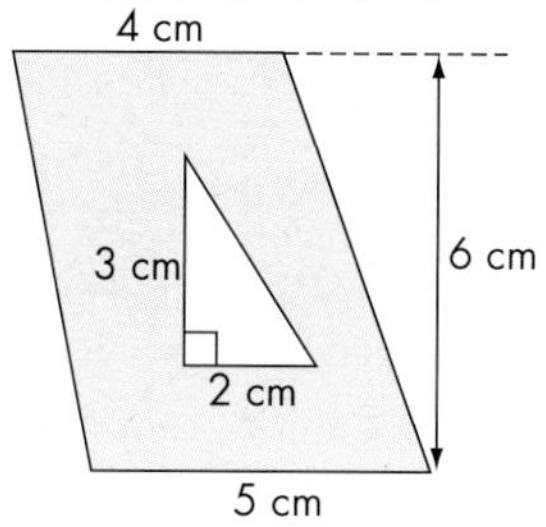

8 What percentage of this shape has been shaded?

7 cm
5 cm
6 cm
3 cm
7 cm
10 cm

9.4 Sectors

Homework 9D

1 For these sectors, calculate:

i the arc length   **ii** the sector area.

a
50°
10 cm

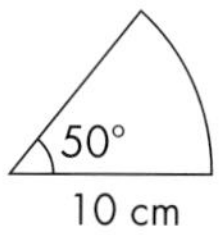

b
90°
7 cm

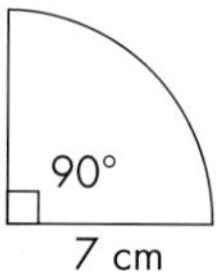

2 Calculate the arc length and the area of a sector whose arc subtends a right angle at the centre of a circle with a diameter of 10 cm. Give your answers in terms of π.

3 Calculate the total perimeter of each shape.

a
20 cm

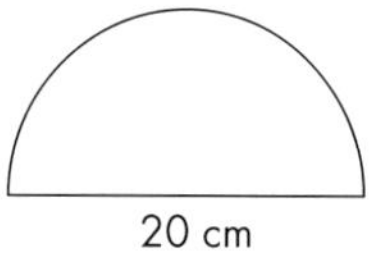

b
12 cm

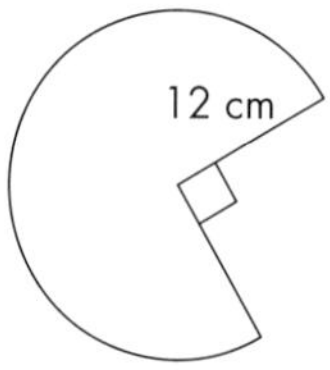

4 Calculate the area of each shape.

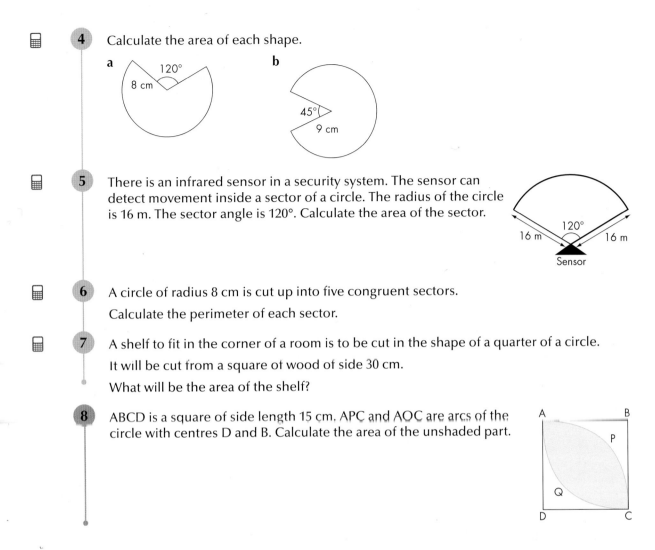

a 120° 8 cm

b 45° 9 cm

5 There is an infrared sensor in a security system. The sensor can detect movement inside a sector of a circle. The radius of the circle is 16 m. The sector angle is 120°. Calculate the area of the sector.

120° 16 m 16 m Sensor

6 A circle of radius 8 cm is cut up into five congruent sectors.

Calculate the perimeter of each sector.

7 A shelf to fit in the corner of a room is to be cut in the shape of a quarter of a circle.

It will be cut from a square of wood of side 30 cm.

What will be the area of the shelf?

8 ABCD is a square of side length 15 cm. APC and AOC are arcs of the circle with centres D and B. Calculate the area of the unshaded part.

A   B
        P
    Q
D   C

9.5 Volume of a prism

Homework 9E

1 Calculate the area of the cross-section and the volume of each prism.

a

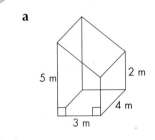

5 m   2 m
    4 m
3 m

b

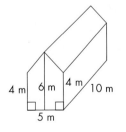

4 m 6 m 4 m 10 m
    5 m

2 A chocolate box is in the shape of a triangular prism. It is 18 cm long and has a volume of 387 cm³.

What is the area of the triangular end of the box?

3 The diagram shows the cross-section of a wooden door wedge.

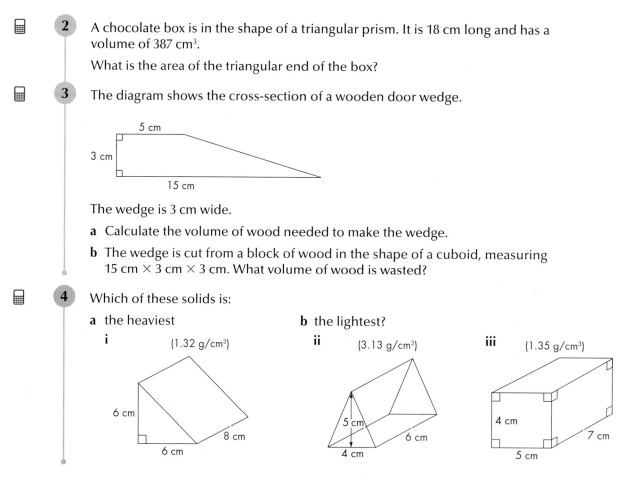

The wedge is 3 cm wide.

a Calculate the volume of wood needed to make the wedge.

b The wedge is cut from a block of wood in the shape of a cuboid, measuring 15 cm × 3 cm × 3 cm. What volume of wood is wasted?

4 Which of these solids is:

a the heaviest **b** the lightest?

i (1.32 g/cm³) **ii** (3.13 g/cm³) **iii** (1.35 g/cm³)

6 cm 6 cm 8 cm 5 cm 4 cm 6 cm 4 cm 5 cm 7 cm

9.6 Cylinders

Homework 9F

1 A cylinder has base radius 5 cm and height 4 cm. Find:

a its volume **b** its curved surface area.

Give your answers in terms of π.

2 A cylinder with base radius 8 cm and height 17 cm. Find:

a its volume **b** its curved surface area.

Give your answers to a suitable degree of accuracy.

3 For the cylinders below, find:

i the volume **ii** the total surface area.

a **b**

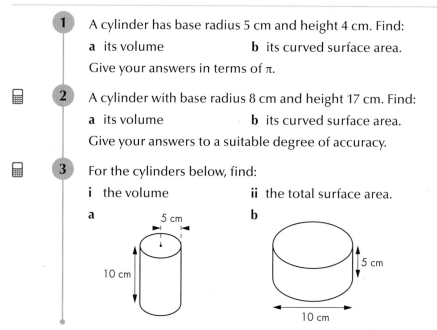

5 cm

10 cm

5 cm

10 cm

4 What is the radius of a cylinder with a height of 6 cm and a volume of 24π cm³?

5 What is the radius of a cylinder with a height of 10 cm and a curved surface area of 360π cm²?

6 What is the height of a cylinder with a diameter of 12 cm and a volume of 108π cm³?

7 A cylinder of height 20 cm has a curved surface area of 200 cm². Calculate the volume of this cylinder.

8 A cylinder has a height of 18 cm and a volume of 390 cm³. Calculate the curved surface area of this cylinder.

9 A cylinder has the same height and radius. The total surface area is 100π. Calculate the volume in terms of π.

10 A square of paper, with side length 10 cm, is bent round to make a cylindrical shape by putting two edges together.

What is the volume of the cylinder?

11 A cylindrical can must have a diameter of 7 cm and a volume of at least 400 cm³.

What is the smallest possible height of the can?

12 Metal cylinders are made by bending rectangular sheets of metal, measuring 15 cm by 6 cm, until the sides meet.

How many cylinders can be made from a sheet of metal that is 2 m long and 1 m wide?

9.7 Volume of a pyramid

Homework 9G

1 Calculate the volume of these rectangular-based pyramids.

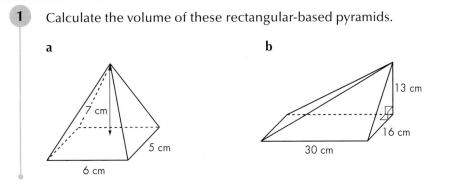

a
7 cm
5 cm
6 cm

b
13 cm
16 cm
30 cm

2　Show that the volume of a pyramid with a square base of side 10 cm and a vertical height of 18 cm is 600 cm³.

3　An octahedron is made by joining two identical square-based pyramids at their bases.

Each pyramid is 9 cm high and has a base with side length 7 cm.

Calculate the volume of the octahedron.

4　The Khufu pyramid in Egypt probably took 20 years to complete and was originally 146 m tall.

It was built from limestone blocks with a density of about 2.7 tonnes per m³.

Each side of the square base was 230 m long.

Estimate the total weight of the blocks used to build the pyramid.

5　Calculate the volume of this shape.

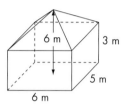

6　Calculate the height h of a rectangular-based pyramid with a length of 14 cm, a width of 10 cm and a volume of 140 cm³.

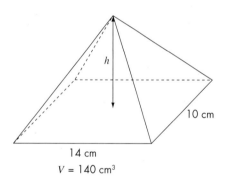

$V = 140$ cm³

7　The pyramid in the diagram has its top 6 cm cut off. The remaining shape is called a frustum. Calculate the volume of the frustum.

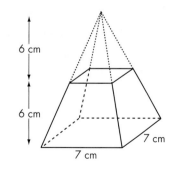

9.8 Cones

Homework 9H 🖩

1 For each cone, calculate:

    i its volume   **ii** its total surface area.

    a   **b**

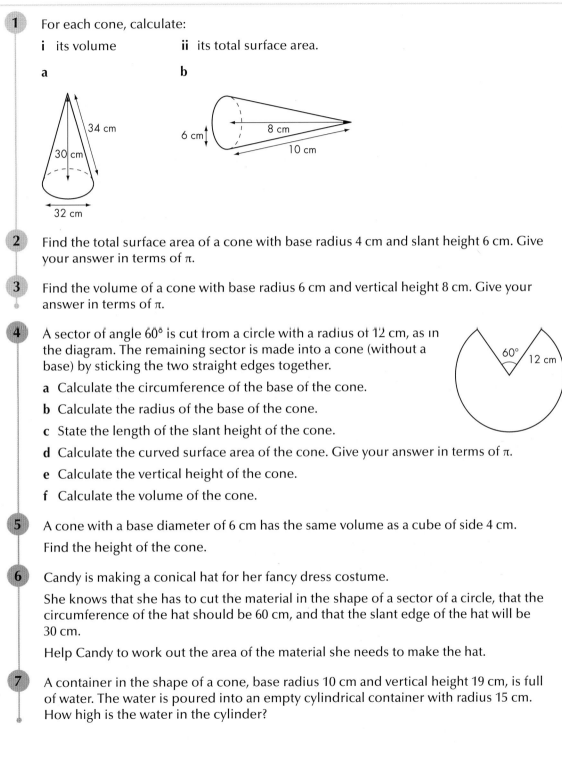

2 Find the total surface area of a cone with base radius 4 cm and slant height 6 cm. Give your answer in terms of π.

3 Find the volume of a cone with base radius 6 cm and vertical height 8 cm. Give your answer in terms of π.

4 A sector of angle 60° is cut from a circle with a radius of 12 cm, as in the diagram. The remaining sector is made into a cone (without a base) by sticking the two straight edges together.

    a Calculate the circumference of the base of the cone.

    b Calculate the radius of the base of the cone.

    c State the length of the slant height of the cone.

    d Calculate the curved surface area of the cone. Give your answer in terms of π.

    e Calculate the vertical height of the cone.

    f Calculate the volume of the cone.

5 A cone with a base diameter of 6 cm has the same volume as a cube of side 4 cm.

    Find the height of the cone.

6 Candy is making a conical hat for her fancy dress costume.

    She knows that she has to cut the material in the shape of a sector of a circle, that the circumference of the hat should be 60 cm, and that the slant edge of the hat will be 30 cm.

    Help Candy to work out the area of the material she needs to make the hat.

7 A container in the shape of a cone, base radius 10 cm and vertical height 19 cm, is full of water. The water is poured into an empty cylindrical container with radius 15 cm. How high is the water in the cylinder?

8 The diagram shows a paper cone. The diameter of the base is 4.8 cm and the slant height is 4 cm. The cone is cut along the line AV and opened out flat, as shown below.

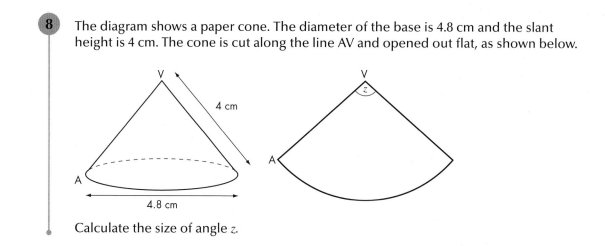

Calculate the size of angle z.

9.9 Spheres

Homework 9I 🖩

1 Calculate the volume of spheres with these measurements. Give your answers in terms of π.

   a Radius 3 cm   **b** Diameter 30 cm

2 Calculate the surface area of spheres with these measurements. Give your answers in terms of π.

   a Radius 4 cm   **b** Diameter 10 cm

3 Calculate the volume and the surface area of a sphere with a diameter of 30 cm.

4 Calculate, correct to one decimal place, the radius of a sphere:

   a with a surface area of 200 cm²   **b** with a volume of 200 cm³.

5 The volume of a sphere is 50 m³. Find its diameter.

6 What is the volume of a sphere with a surface area of 400 cm²?

7 A hemispherical hole of diameter 4 cm is cut out of a metal cube of side 5 cm.

   What is the volume of the resulting shape?

8 A manufacturing company makes steel ball bearings with a diameter of 4 mm for their roller skates. How many ball bearings can the company make from 1 m³ of steel?

9 The diagram shows a spinning top made from a cone with base radius 6 cm and slant height 10 cm, and a hemisphere of radius 6 cm.

   a Calculate the volume of the spinning top.

   b Calculate the total surface area of the spinning top.

   Give your answers in terms of π.

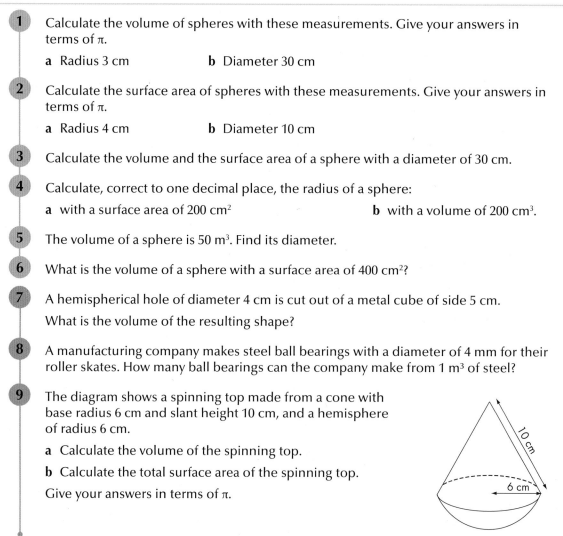

10 Algebra: Linear graphs

10.1 Drawing linear graphs from points

Homework 10A

Follow these steps when drawing graphs.

Step 1: Use the highest and lowest values of x given in the range.

Step 2: If the first part of the function is a division, pick x-values that divide exactly to avoid fractions.

Step 3: Always label your graph with its equation. This is particularly important when you are drawing two graphs on the same set of axes.

Step 4: Create a table of values.

1. Draw the graph of $y = 2x + 3$ for x-values from 0 to 5 ($0 \leqslant x \leqslant 5$).

2. Draw the graph of $y = 3x - 1$ for $0 \leqslant x \leqslant 5$.

3. Draw the graph of $y = \frac{x}{2} - 2$ for $0 \leqslant x \leqslant 12$.

4. Draw the graph of $y = 2x + 1$ for $-2 \leqslant x \leqslant 2$.

5. Draw the graph of $y = \frac{x}{2} + 5$ for $-6 \leqslant x \leqslant 6$.

6. a On the same set of axes, draw the graphs of $y = 3x - 1$ and $y = 2x + 3$ for $0 \leqslant x \leqslant 5$.
   b At which point do the two lines intersect?

7. a On the same axes, draw the graphs of $y = 4x - 3$ and $y = 3x + 2$ for $0 \leqslant x \leqslant 6$.
   b At which point do the two lines intersect?

8. a On the same axes, draw the graphs of $y = \frac{x}{2} + 1$ and $y = \frac{x}{3} + 2$ for $0 \leqslant x \leqslant 12$.
   b At which point do the two lines intersect?

9. a On the same axes, draw the graphs of $y = 2x + 3$ and $y = 2x - 1$ for $0 \leqslant x \leqslant 4$.
   b Do the graphs intersect? If not, explain why.

10. a Copy and complete the table of values for the equation $x + y = 6$.

x	0	1	2	3	4	5	6
y							

   b Now draw the graph of $x + y = 3$.

11 CityCabs uses this formula to work out the cost (£) of a journey of k kilometres:

$C = 2.5 + k$

TownCars uses this formula to work out the cost of a journey of k kilometres:

$C = 2 + 1.25k$

a On a copy of the grid, draw lines to represent these formulae.

b At what length of journey do CityCabs and TownCars charge the same amount?

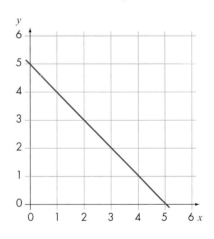

12 The diagram shows the graph of $x + y = 5$.

Draw a line of the form $x = a$ and a line of the form $y = b$ so that the area between the 3 lines is 4.5 square units.

10.2 Gradient of a line

Homework 10B

1 Draw lines with these gradients.

a 3 **b** $\frac{1}{2}$ **c** −1 **d** 8 **e** $\frac{3}{4}$ **f** $-\frac{1}{3}$

2 **a** Draw a pair of axes with x and y from −10 to 10.

b Draw one line with each of these gradients, starting from the origin each time. Remember to label each line.

i $\frac{1}{2}$ **ii** 1 **iii** 2 **iv** 4

v −4 **vi** −2 **vii** −1 **viii** $-\frac{1}{2}$

c Describe the symmetries of your diagram.

3 Find the gradient of lines **a** to **j**.

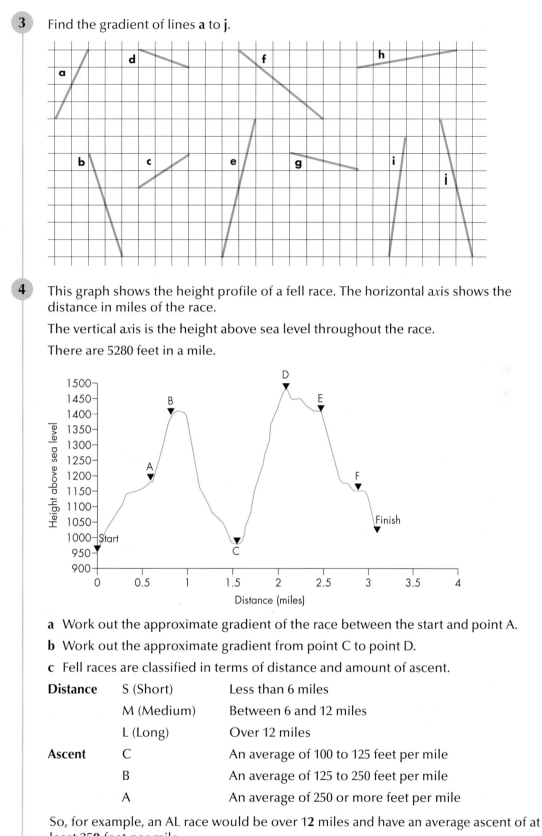

4 This graph shows the height profile of a fell race. The horizontal axis shows the distance in miles of the race.

The vertical axis is the height above sea level throughout the race.

There are 5280 feet in a mile.

a Work out the approximate gradient of the race between the start and point A.

b Work out the approximate gradient from point C to point D.

c Fell races are classified in terms of distance and amount of ascent.

Distance	S (Short)	Less than 6 miles
	M (Medium)	Between 6 and 12 miles
	L (Long)	Over 12 miles
Ascent	C	An average of 100 to 125 feet per mile
	B	An average of 125 to 250 feet per mile
	A	An average of 250 or more feet per mile

So, for example, an AL race would be over **12** miles and have an average ascent of at least **250** feet per mile.

In which category is the race shown in the graph?

5 Write the gradients of these lines in the form 1 : n.

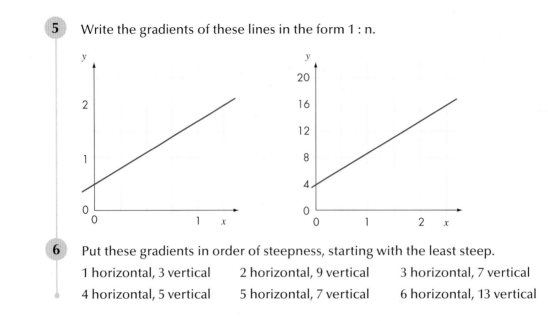

6 Put these gradients in order of steepness, starting with the least steep.

1 horizontal, 3 vertical   2 horizontal, 9 vertical   3 horizontal, 7 vertical

4 horizontal, 5 vertical   5 horizontal, 7 vertical   6 horizontal, 13 vertical

10.3 Drawing graphs by gradient-intercept and cover-up methods

Homework 10C

1 Draw these lines using the gradient-intercept method. Use the same grid, taking both x and y from −10 to 10. If the grid gets too 'crowded', draw another one.

a $y = 2x + 4$ **b** $y = 3x - 2$ **c** $y = x + 1$ **d** $y = x - 1$

e $y = 6x - 2$ **f** $y = x + 3$ **g** $y = x - 2$ **h** $y = 3x - 4$

For questions **2** and **3**, draw grids with $-6 \leqslant x \leqslant 6$ and $-8 \leqslant y \leqslant 8$.

2 **a** Use the gradient-intercept method to draw these lines on the same grid.

    i $y = 3x + 2$ **ii** $y = 2x - 1$

    b Where do the lines cross?

3 Here are the equations of three lines.

A: $y = 4x - 3$ B: $2y = 8x - 6$ C: $y = 2x - 3$

    a State a mathematical property that lines A and B have in common.

    b State a mathematical property that lines B and C have in common.

    c Which of the following points is the intersection of lines A and C?

       (1, −3)   (0, 3)   (0, −3)   (1, 3)

4 a What is the gradient of line A?

b What is the gradient of line B?

c What angle is there between line A and B?

d What is the relationship between the gradients of lines A and B?

e Another line, C, has a gradient of $-\frac{1}{2}$.

What is the gradient of a line perpendicular to C?

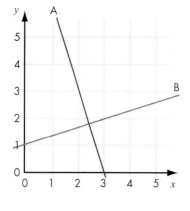

Homework 10D

1 Draw these lines using the cover-up method. Use the same grid, taking both x and y from −10 to 10. If the grid gets too 'crowded', draw another.

a $2x + 3y = 6$ b $3x + 4y = 12$ c $5x − 4y = 20$ d $x + y = 8$

e $2x − 3y = 18$ f $x − y = 6$ g $3x − 5y = 15$ h $3x − 2y = 12$

i $5x + 4y = 30$ j $x + y = −1$ k $x + y = 5$ l $x − y = −6$

2 a Use the cover-up method to draw these lines on the same grid.

i $x + 2y = 4$ ii $2x − y = 2$

b Where do the lines cross?

3 a Use the cover-up method to draw these lines on the same grid.

i $x + 2y = 6$ ii $2x − y = 2$

b Where do the lines cross?

4 Here are the equations of three lines.

A: $3x + 4y = 12$ B: $x − 2y = 3$ C: $x + y = 3$

a State a mathematical property that lines A and C have in common.

b State a mathematical property that lines B and C have in common.

c Line A crosses the x axis at (4, 0). Line B crosses the y axis at $(0, -1\frac{1}{2})$.

Find values of a and b such that the line $ax − by = 12$ passes through (4, 0) and $(0, -1\frac{1}{2})$.

5 The diagram shows a hexagon ABCDEF.

The equation of the line through A and B is $y = 3$.

The equation of the line through B and C is $x + y = 4$

a Write down the equation of the lines through these vertices.

i E and D

ii D and C

iii A and F

iv E and F

b The gradient of the line through E and B is 2.

Write down the gradient of the line through A and D.

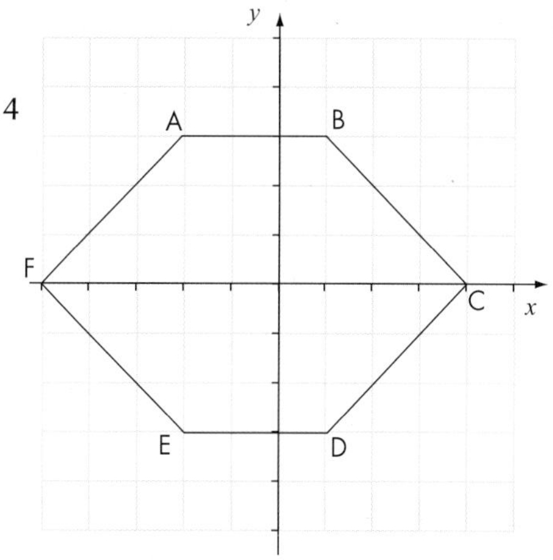

10.4 Finding the equation of a line from its graph

Homework 10E

1 In each grid, there are two lines.

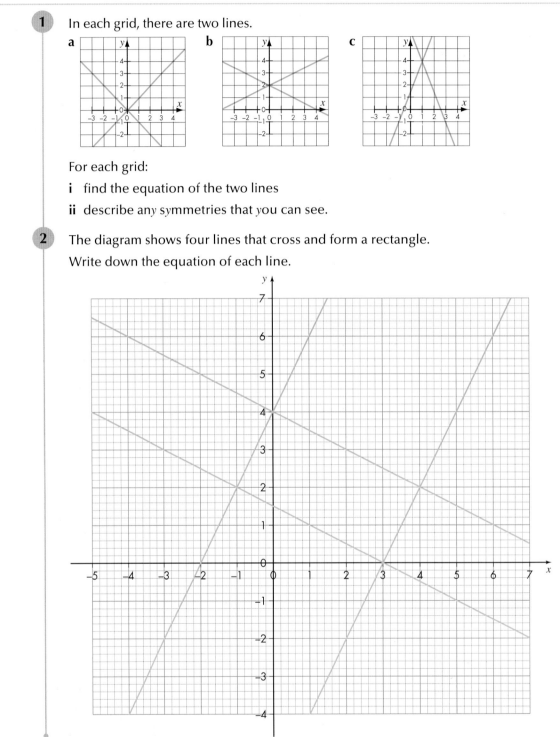

For each grid:

i find the equation of the two lines

ii describe any symmetries that you can see.

2 The diagram shows four lines that cross and form a rectangle.

Write down the equation of each line.

3 A straight line passes through the points (–1, 3) and (–2, 5).

    a Describe how you can tell that the line also passes through (0, 1).

    b Describe how you can tell that the line has a gradient of –2.

    c Work out the equation of the line that passes through (–1, 5) and (–2, 8).

4 A line passes through the points (4, 0) and (0, 12). Find the equation of the line in the form $ax + by = c$.

5 Line L1 passes through the points (4, 6) and (1, 3).

    Line L2 passes through the points (0, –4) and (2, 2).

    Find the coordinates of the point where the two lines intersect.

6 Kate has drawn a straight-line graph that passes through (1, –1) and (3, 3).

    Richard says that the point (4, 5) will lie on the graph.

    Rachel says that the point (4, 5) will not lie on the graph.

    Who is correct? Explain your answer.

10.5 Real-life uses of graphs

Homework 10F

1 This graph shows the charges made by an electricity company.

    a The standing charge is the basic charge before the cost per unit is added. What is the standing charge?

    b What is the gradient of the line?

    c Write down the rule to work out the total charge for different amounts of electricity.

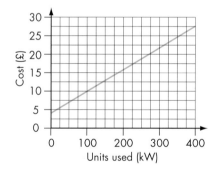

2 This graph shows the charges made by a gas company.

    a What is the standing charge?

    b What is the gradient of the line?

    c Write down the rule to work out the total charge for different amounts of gas.

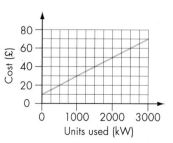

3 This graph illustrates the charges made by a phone company.

    a What is the standing charge?

    b What is the gradient of the line?

    c Write down the rule to calculate the total charge with this phone company.

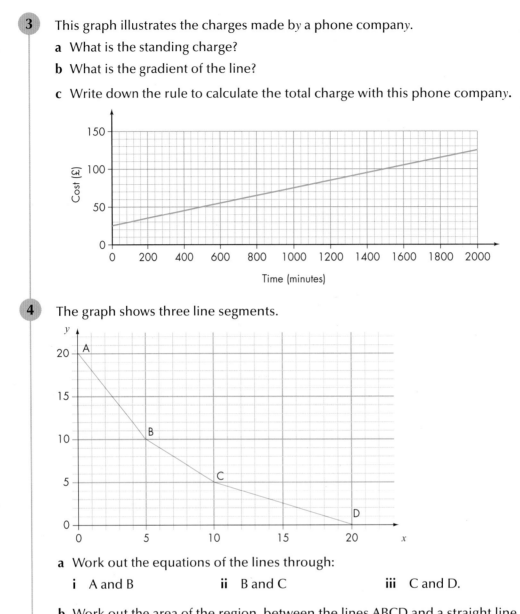

4 The graph shows three line segments.

    a Work out the equations of the lines through:

      i A and B   **ii** B and C   **iii** C and D.

    b Work out the area of the region between the lines ABCD and a straight line between A and D.

10.6 Solving simultaneous equations using graphs

Homework 10G

Draw the graphs to find the solution of each pair of simultaneous equations.

1

  a $x + 4y = 1$
       $x - y = 6$

  b $y = 2x + 1$
       $3x + 2y = 23$

  c $y = 2x + 5$
       $y = x + 4$

  d $y = x$
       $x - y = 4$

  e $y + 10 = 2x$
       $5x + y = 18$

  f $y = 5x - 1$
       $y = 3x + 2$

  g $y = x + 11$
       $x + y = 5$

  h $y - 3x = 8$
       $y = x + 6$

  i $y = -x$
       $y = 4x + 15$

  j $3x + 2y = 2$
       $y = -2x$

  k $y = 3x - 4$
       $y + x = 6$

  l $y = 3x - 12$
       $x + y = 2$

2 Two coffees and three cakes cost £7.00.

Two coffees and one cake cost £4.00.

Use x to represent the price of a coffee and y to represent the cost of a cake.

Use graphs to write down the cost of a coffee and the cost of a cake.

3 The graph shows four lines.

P: $y = -x$ Q: $y = 2x + 6$ R: $y = x - 2$ S: $y = -\frac{2}{3}x - 2$

a Which pairs of lines intersect at the following points?

   i $(0, -2)$ **ii** $(-3, 0)$ **iii** $(1, -1)$ **iv** $(-2, 2)$

b Solve the simultaneous equations given by P and S to find the point of intersection of these lines.

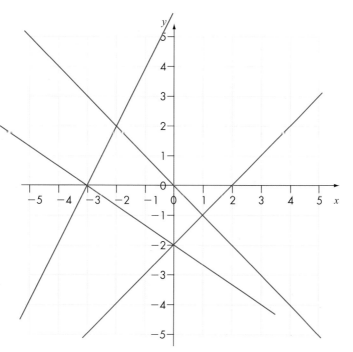

10.7 Parallel and perpendicular lines

Homework 10H

1 Two taxi companies, Acabs and BeeCabs, use the same graph to charge for journeys. A sketch of this graph is shown here.

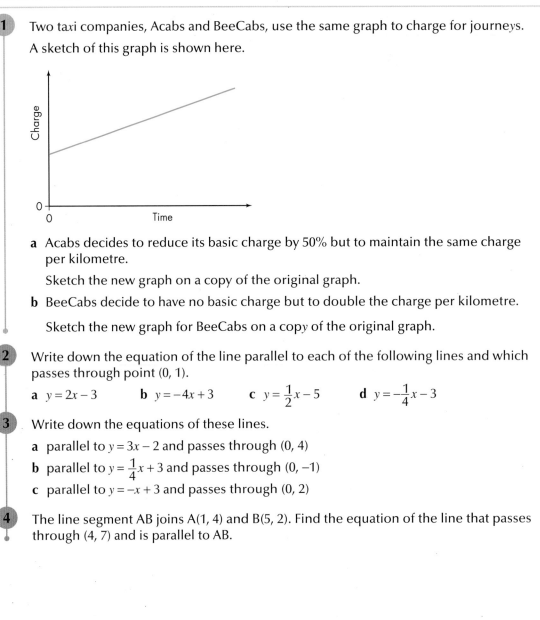

   a Acabs decides to reduce its basic charge by 50% but to maintain the same charge per kilometre.

   Sketch the new graph on a copy of the original graph.

   b BeeCabs decide to have no basic charge but to double the charge per kilometre.

   Sketch the new graph for BeeCabs on a copy of the original graph.

2 Write down the equation of the line parallel to each of the following lines and which passes through point (0, 1).

   a $y = 2x - 3$ **b** $y = -4x + 3$ **c** $y = \frac{1}{2}x - 5$ **d** $y = -\frac{1}{4}x - 3$

3 Write down the equations of these lines.

   a parallel to $y = 3x - 2$ and passes through (0, 4)

   b parallel to $y = \frac{1}{4}x + 3$ and passes through (0, −1)

   c parallel to $y = -x + 3$ and passes through (0, 2)

4 The line segment AB joins A(1, 4) and B(5, 2). Find the equation of the line that passes through (4, 7) and is parallel to AB.

11 Geometry: Right-angled triangles

11.1 Pythagoras' theorem

Homework 11A

1 Calculate the length of the hypotenuse for each triangle. Give your answers to a suitable degree of accuracy.

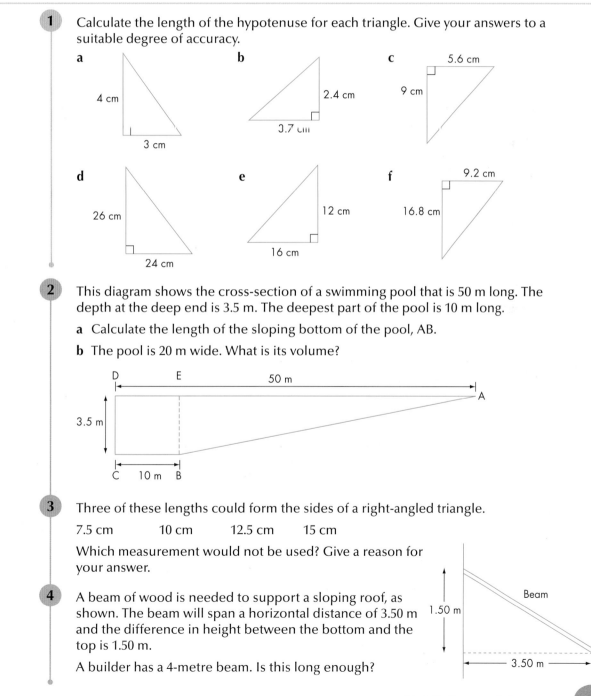

a 4 cm, 3 cm

b 2.4 cm, 3.7 cm

c 5.6 cm, 9 cm

d 26 cm, 24 cm

e 12 cm, 16 cm

f 9.2 cm, 16.8 cm

2 This diagram shows the cross-section of a swimming pool that is 50 m long. The depth at the deep end is 3.5 m. The deepest part of the pool is 10 m long.

a Calculate the length of the sloping bottom of the pool, AB.

b The pool is 20 m wide. What is its volume?

D   E   50 m   A

3.5 m

C   10 m   B

3 Three of these lengths could form the sides of a right-angled triangle.

7.5 cm   10 cm   12.5 cm   15 cm

Which measurement would not be used? Give a reason for your answer.

4 A beam of wood is needed to support a sloping roof, as shown. The beam will span a horizontal distance of 3.50 m and the difference in height between the bottom and the top is 1.50 m.

A builder has a 4-metre beam. Is this long enough?

Beam

1.50 m

3.50 m

11.2 Finding the length of a shorter side

Homework 11B

1 Calculate the length of x for each triangle. Give your answers to a suitable degree of accuracy.

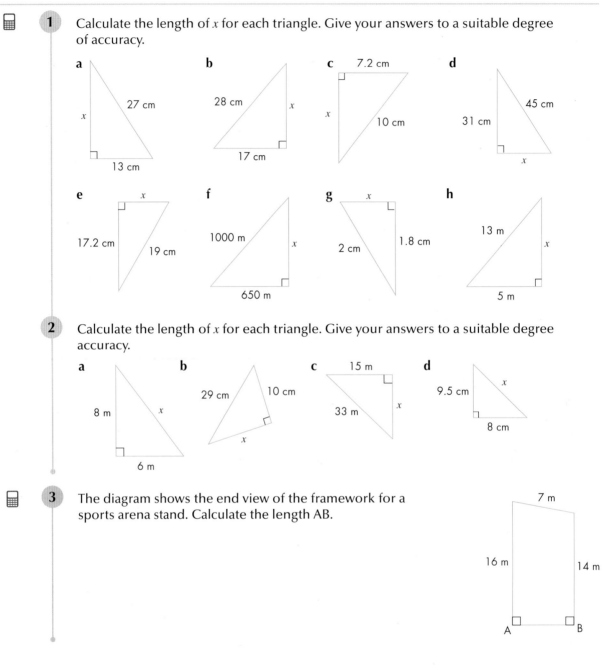

2 Calculate the length of x for each triangle. Give your answers to a suitable degree accuracy.

3 The diagram shows the end view of the framework for a sports arena stand. Calculate the length AB.

4 Calculate the lengths of a and b.

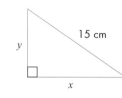

5 The lengths of the three sides of a right-angled triangle are all integer values. The hypotenuse is 15 cm. How long are the two other sides?

11.3 Applying Pythagoras' theorem in real-life situations

Homework 11C

1 A ladder 3.8 m long is placed against a wall. The foot of the ladder is 1.1 m from the wall.

A window cleaner can reach windows that are 1 m above the top of her ladder. Can she reach a window that is 4 m above the ground?

2 A ladder, 15 m long, leans against a wall. The ladder must reach 12 m up the wall. How far away from the wall should the foot of the ladder be placed?

3 A rectangle is 3 m long and 1.2 m wide. How long is the diagonal?

4 How long is the diagonal of a square with a side of 10 m?

5 A ship going from a port to a lighthouse steams 8 km east and 6 km north. How far is the lighthouse from the port?

6 The diagram shows three towns, A, B and C, joined by two roads. The council wants to build a road that runs directly from A to C.

How much shorter will the new road be than the two existing roads?

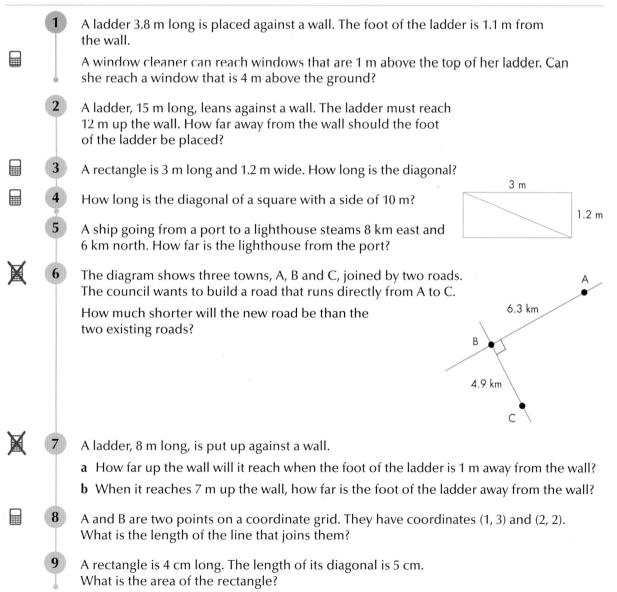

7 A ladder, 8 m long, is put up against a wall.

a How far up the wall will it reach when the foot of the ladder is 1 m away from the wall?

b When it reaches 7 m up the wall, how far is the foot of the ladder away from the wall?

8 A and B are two points on a coordinate grid. They have coordinates (1, 3) and (2, 2). What is the length of the line that joins them?

9 A rectangle is 4 cm long. The length of its diagonal is 5 cm. What is the area of the rectangle?

10 Is the triangle with sides 11 cm, 60 cm and 61 cm a right-angled triangle?
Give a reason for your answer.

11 A and B are two points on a coordinate grid. They have coordinates (−3, −7) and (4, 6).
Show that the line that joins them has length 14.8 units.

12 A boat sails due east from A for 27 km. It then changes course
and sails due south for 30 km. On a map, the distance between
A and C is 10.8 cm.

a What is the scale of the map?

b What is the direct distance from A to B in kilometres?

A ——27 km—— C   N

30 km

Not to scale

B

13 A mobile phone mast is supported by a cable attached to the ground from the top of
the mast. The mast is 12.5 m high and the cable is 17.8 m long.

How far from the foot of the mast is the cable fixed to the ground?

14 A rolling pin is 45 cm long.

Will it fit inside a kitchen drawer with internal measurements of 40 cm by 33 cm?

Give reasons for your answer.

11.4 Pythagoras' theorem and isosceles triangles

Homework 11D

1 Calculate the area of these isosceles triangles.

10 cm   10 cm   5 cm

7 cm   4 cm

2 Calculate the area of an isosceles triangle with sides of 10 cm, 10 cm and 8 cm.

3 Calculate the area of an equilateral triangle of side 10 cm.

4 **a** Calculate the area of an equilateral triangle with sides of 20 cm.

b Show that the answer to part a is not double the answer to question **3**.

5 An isosceles triangle has sides of 6 cm and 8 cm.

a Sketch the two isosceles triangles that fit this data.

b Which of the two triangles has the greater area?

6 Calculate the area of this isosceles triangle.

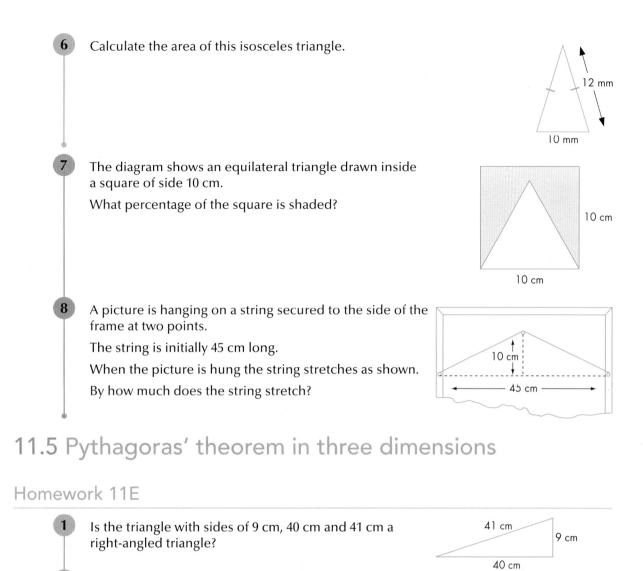

12 mm

10 mm

7 The diagram shows an equilateral triangle drawn inside a square of side 10 cm.

What percentage of the square is shaded?

10 cm

10 cm

8 A picture is hanging on a string secured to the side of the frame at two points.

The string is initially 45 cm long.

When the picture is hung the string stretches as shown.

By how much does the string stretch?

10 cm

45 cm

11.5 Pythagoras' theorem in three dimensions

Homework 11E

1 Is the triangle with sides of 9 cm, 40 cm and 41 cm a right-angled triangle?

41 cm

9 cm

40 cm

2 A garage is 5 m long, 5 m wide and 2 m high. Can a pole 7 m long be stored in it?

3 Spike, a spider, is at the corner S of the wedge shown in the diagram. Fred, a fly, is at the corner F of the same wedge.

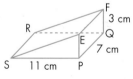

a If Spike only uses the edges of the wedge, her journey to Fred will be of two possible lengths. Calculate both possible distances.

F
3 cm
R
Q
E
7 cm
S   11 cm   P

b Calculate the shortest distance Spike would have to travel across the face of the wedge to get directly to Fred.

4 A corridor is 5 m wide and turns through a right angle, as in the diagram.

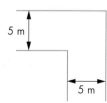

a What is the longest pole that can be carried along the corridor horizontally?

5 m

b If the corridor is 3 m high, what is the longest pole that can be carried along in any direction?

5 m

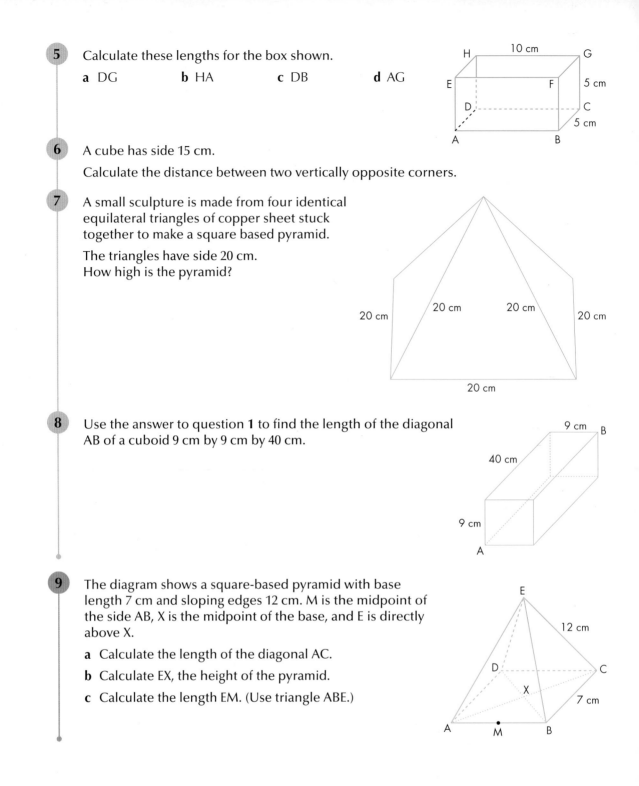

5 Calculate these lengths for the box shown.

    a DG   **b** HA   **c** DB   **d** AG

6 A cube has side 15 cm.

    Calculate the distance between two vertically opposite corners.

7 A small sculpture is made from four identical equilateral triangles of copper sheet stuck together to make a square based pyramid.

    The triangles have side 20 cm.
How high is the pyramid?

8 Use the answer to question **1** to find the length of the diagonal AB of a cuboid 9 cm by 9 cm by 40 cm.

9 The diagram shows a square-based pyramid with base length 7 cm and sloping edges 12 cm. M is the midpoint of the side AB, X is the midpoint of the base, and E is directly above X.

    a Calculate the length of the diagonal AC.

    b Calculate EX, the height of the pyramid.

    c Calculate the length EM. (Use triangle ABE.)

11.6 Trigonometric ratios

Homework 11F

In these questions, give any answers involving angles to the nearest degree.

1 Find these values, rounding your answers to 3 significant figures.

    a $\sin 52°$ **b** $\sin 46°$ **c** $\sin 76.3°$ **d** $\sin 90°$

2 Find these values, rounding your answers to 3 significant figures.

    a $\cos 52°$ **b** $\cos 46°$ **c** $\cos 76.3°$ **d** $\cos 90°$

3 **a** Calculate $(\sin 52°)^2 + (\cos 52°)^2$.   **b** Calculate $(\sin 46°)^2 + (\cos 46°)^2$.

    c Calculate $(\sin 76.3°)^2 + (\cos 76.3°)^2$.   **d** Calculate $(\sin 90°)^2 + (\cos 90°)^2$.

    e What do you notice about your answers?

4 Use your calculator to work out the value of:

    a $\tan 52°$ **b** $\tan 46°$ **c** $\tan 76.3°$ **d** $\tan 0°$

5 Use your calculator to find these values.

    a $\sin 52° \div \cos 52°$ **b** $\sin 46° \div \cos 46°$ **c** $\sin 76.3° \div \cos 76.3°$

    d $\sin 0° \div \cos 0°$ **e** What connects these answers to the answers to question **4**?

6 Use your calculator to work out these values.

    a $6 \sin 55°$ **b** $7 \cos 45°$ **c** $13 \sin 67°$ **d** $20 \tan 38°$

7 Use your calculator to work out these values.

    a $\dfrac{6}{\sin 55°}$ **b** $\dfrac{7}{\cos 45°}$ **c** $\dfrac{13}{\sin 67°}$ **d** $\dfrac{20}{\tan 38°}$

8 Calculate $\sin x$, $\cos x$ and $\tan x$ for this triangle. Leave your answers as fractions.

9 You are told that $\sin x = \dfrac{5}{\sqrt{34}}$. Show how you can find $\tan x$ from this information without using a calculator.

11.7 Calculating angles

Homework 11G

Use your calculator to find the answers to the following. Give your answers to 1 decimal place.

1 What angles have the following sines?

    a 0.4   **b** 0.707   **c** 0.879   **d** 0.666666666666666…

2 What angles have the following cosines?

    a 0.4   **b** 0.707   **c** 0.879   **d** 0.333333333333333…

3 What angles have the following tangents?

    a 0.4   **b** 1.24   **c** 0.875   **d** 2.625

4 What angles have the following sines?

    a $3 \div 8$ **b** $1 \div 3$ **c** $3 \div 10$ **d** $5 \div 8$

5 What angles have the following cosines?

    a $3 \div 8$ **b** $1 \div 3$ **c** $3 \div 10$ **d** $5 \div 8$

6 What angles have the following tangents?

    a $3 \div 8$ **b** $3 \div 2$ **c** $5 \div 7$ **d** $19 \div 5$

7 If sin 54° = 0.809 to 3 decimal places, what angle has a cosine of 0.809?

11.8 Using the sine and cosine functions

Homework 11H

1 Calculate the value of x in each of these triangles.

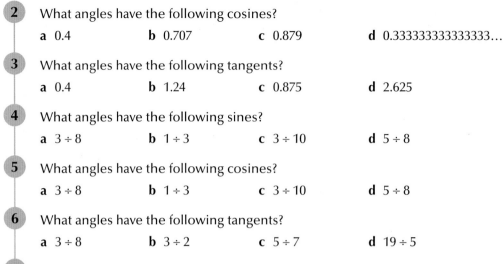

2 Angle θ has a sine of $\frac{7}{20}$. Calculate the lengths marked x in these triangles.

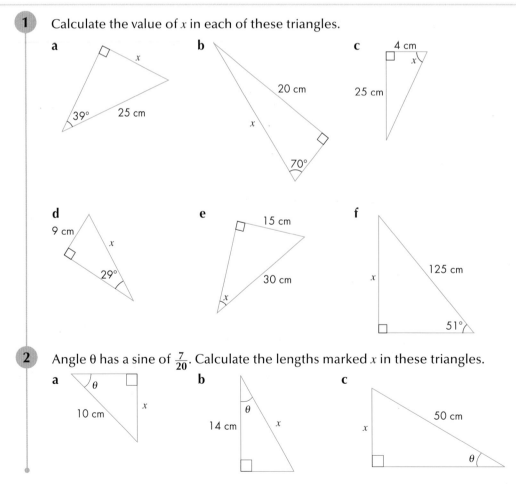

3 Caxton is due north of Ashville and due west of Peaton. A pilot flies directly from Ashville to Peaton, a distance of 15 km, on a bearing of 050°.

a Calculate the direct distance from Caxton to Peaton.

b Work out the bearing of Ashville from Peaton.

Homework 11I

1 Calculate the value of x in each of these triangles.

a

b

c

d

e

f

2 Angle θ has a cosine of $\frac{7}{15}$. Calculate the lengths marked x in these triangles.

a

b

c

3 The diagram shows the positions of three telephone masts, A, B and C.

Mast C is 6 km due east of Mast B.

Mast A is due north of Mast B, and 9 km from Mast C.

a Calculate the direct distance between A and B.

Give your answer in kilometres, correct to 3 significant figures.

b Calculate the size of the angle marked x.

Give your angle correct to 1 decimal place.

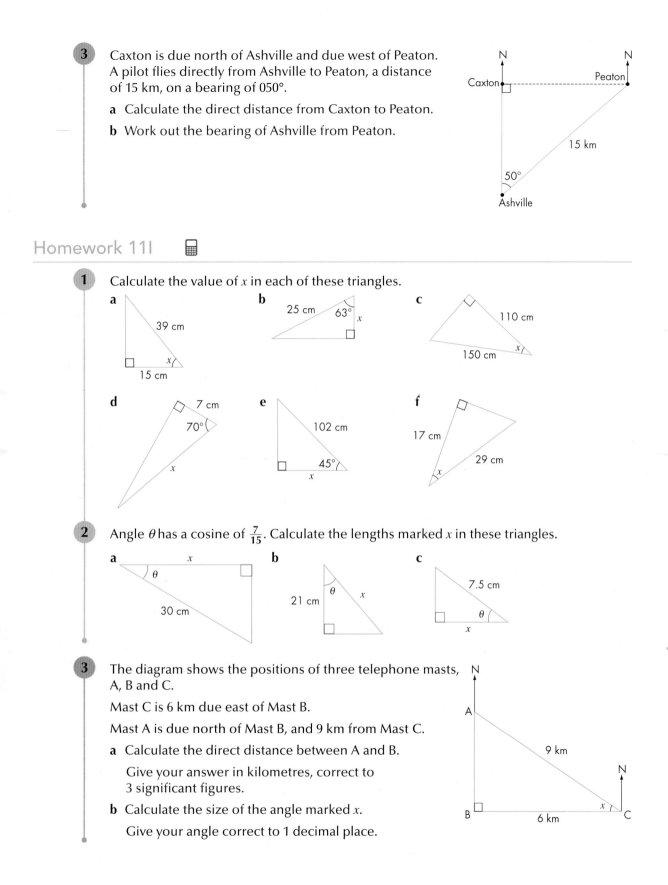

11.9 Using the tangent function

Homework 11J 🖩

1 Calculate the value of x in each triangle.

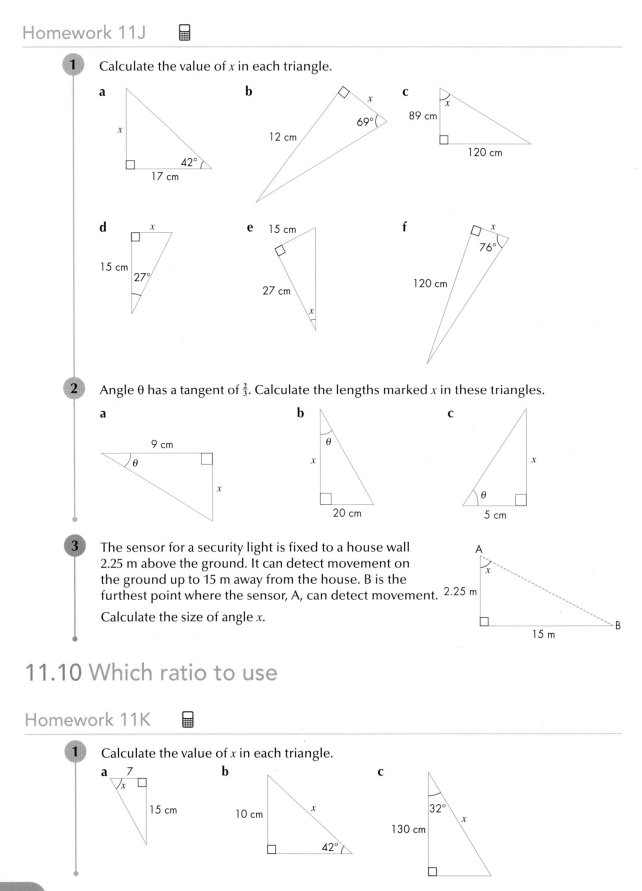

a

x

17 cm

42°

b

12 cm

69°

x

c

89 cm

x

120 cm

d

x

15 cm

27°

e

15 cm

27 cm

x

f

x

76°

120 cm

2 Angle θ has a tangent of $\frac{2}{3}$. Calculate the lengths marked x in these triangles.

a

9 cm

θ

x

b

θ

x

20 cm

c

x

θ

5 cm

3 The sensor for a security light is fixed to a house wall 2.25 m above the ground. It can detect movement on the ground up to 15 m away from the house. B is the furthest point where the sensor, A, can detect movement.

Calculate the size of angle x.

A

x

2.25 m

15 m

B

11.10 Which ratio to use

Homework 11K 🖩

1 Calculate the value of x in each triangle.

a

7

x

15 cm

b

10 cm

x

42°

c

32°

x

130 cm

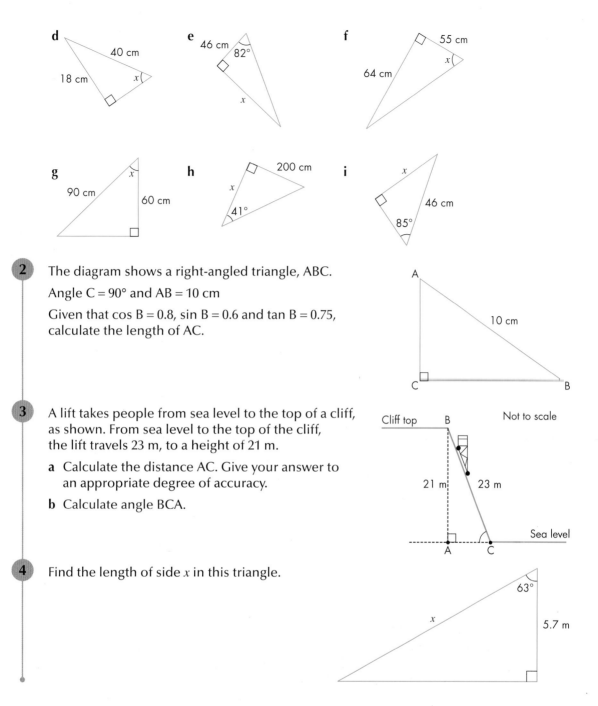

d 40 cm 18 cm x

e 46 cm 82° x

f 55 cm 64 cm x

g x 90 cm 60 cm

h 200 cm x 41°

i x 46 cm 85°

2 The diagram shows a right-angled triangle, ABC.

Angle C = 90° and AB = 10 cm

Given that cos B = 0.8, sin B = 0.6 and tan B = 0.75, calculate the length of AC.

A 10 cm C B

3 A lift takes people from sea level to the top of a cliff, as shown. From sea level to the top of the cliff, the lift travels 23 m, to a height of 21 m.

a Calculate the distance AC. Give your answer to an appropriate degree of accuracy.

b Calculate angle BCA.

Cliff top B Not to scale 21 m 23 m Sea level A C

4 Find the length of side x in this triangle.

63° 5.7 m x

11.11 Solving problems using trigonometry

Homework 11L

In these questions, give any answers involving angles to the nearest degree.

1 A ladder, 8 m long, rests against a wall. The foot of the ladder is 2.7 m from the base of the wall. What angle does the ladder make with the ground?

2 The ladder in Question **1** has a 'safe angle' with the ground of between 70° and 80°. What are the safe limits for the distance of the foot of the ladder from the wall? How high up the wall does the ladder reach?

3 Angela paces out 60 m from the base of a block of flats, She then measures the angle to the top of the flats as 42°. How would Alicia find the height of the block of flats?

42°
60 m

4 A slide makes an angle of 46° with the ground. The slide is 7 m long. How high above the ground is the top of the slide?

7 m
46°

5 A rectangle measures 9 cm by 5 cm. Use trigonometry to calculate the angle that the diagonal makes with the long side of the rectangle.

6 Drumsbury Town Council wants to put up a flag pole outside the town hall. The diagram shows the end view of the town hall building.

Regulations state that the flag pole must not be more than half the height of the building.

What is the maximum height that the flag pole can be?

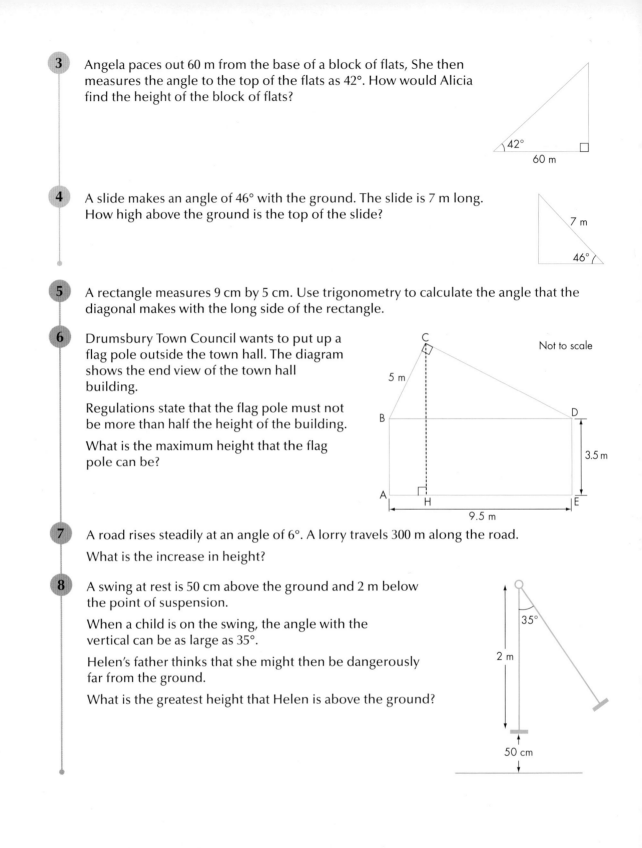

Not to scale

C
5 m
B
A
H
9.5 m
D
3.5 m
E

7 A road rises steadily at an angle of 6°. A lorry travels 300 m along the road.

What is the increase in height?

8 A swing at rest is 50 cm above the ground and 2 m below the point of suspension.

When a child is on the swing, the angle with the vertical can be as large as 35°.

Helen's father thinks that she might then be dangerously far from the ground.

What is the greatest height that Helen is above the ground?

35°
2 m
50 cm

Homework 11M 🖩

In these questions, give any answers involving angles to the nearest degree.

1 Ric sees an aircraft in the sky. The aircraft is at a horizontal distance of 15 km from Ric. The angle of elevation is 42°. How high is the aircraft?

2 A man standing 100 m from the base of a block of flats, looks at the top of the block and notices that the angle of elevation is 49°. How high is the block of flats?

3 A man stands 15 m from a tree. The angle of elevation of the top of the tree from his eye is 25°. If his eye is 1.5 m above the ground, how tall is the tree?

4 A bird sitting at the very top of the tree in question **3**, sees a worm next to the foot of the man. What is the angle of depression from the bird's eye to the worm?

5 I walk 200 m away from a chimney that is 120 m high. What is the angle of elevation from my eye to the top of the chimney? (Ignore the height of my eye above the ground.)

6 The height my eye above the ground in question **5** is 1.8 m. What is the difference in the angle of elevation?

7 A boat is moored 50 m from the foot of a vertical cliff, at B. The angle of depression of the boat from the top of the cliff is 52°.

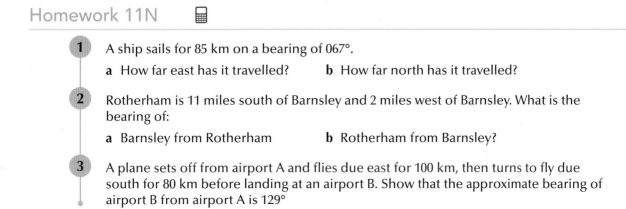

    a Calculate the height of the cliff.

    b The boat is released from its mooring and it drifts 350 m further away from the cliff. Calculate the angle of elevation of the top of the cliff from the boat.

8 A boat is 450 m from the base of a cliff. The angle of elevation of the top of the cliff is 8°.

How high is the cliff?

9 To find the height of a tree, Sacha tries to measure the angle of elevation of the top from a point 40 m away.

He finds it difficult to measure the angle accurately, but thinks it is between 30° and 35°.

What can you tell him about the height of the tree?

11.12 Trigonometry and bearings

Homework 11N 🖩

1 A ship sails for 85 km on a bearing of 067°.

    a How far east has it travelled?   **b** How far north has it travelled?

2 Rotherham is 11 miles south of Barnsley and 2 miles west of Barnsley. What is the bearing of:

    a Barnsley from Rotherham   **b** Rotherham from Barnsley?

3 A plane sets off from airport A and flies due east for 100 km, then turns to fly due south for 80 km before landing at an airport B. Show that the approximate bearing of airport B from airport A is 129°

4 Mountain A is due east of a walker. Mountain B is due south of the same walker. The guidebook says that mountain A is 5 km from mountain B, on a bearing of 038°. How far is the walker from mountain B?

5 The diagram shows the relative distances and bearings of three ships A, B and C.

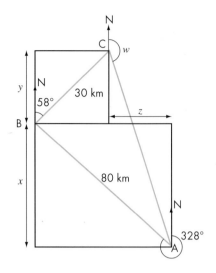

   a How far north of A is B?
      (Distance x on diagram.)

   b How far north of B is C?
      (Distance y on diagram.)

   c How far west of A is C?
      (Distance z on diagram.)

   d What is the bearing of A from C?
      (Angle w on diagram.)

6 A plane is flying from Leeds (L) to London Heathrow (H).

It flies 150 miles on a bearing of 136° to point A. It then turns through 90° and flies the final 80 miles to H.

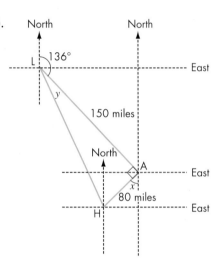

   a i Show clearly why the angle marked x is 46°.

     ii Give the bearing of H from A.

   b Use Pythagoras' theorem to calculate the distance LH.

   c i Calculate the size of the angle marked y.

     ii Work out the bearing of L from H.

7 A plane flies 200 km on a bearing of 124° and then 150 km on a bearing of 053°.

How far east has it flown?

8 Large boats are supposed to stay at least 300 m from the shore at Longup Beach.

Don notices a large boat that is due north of his position on the beach. He walks 100 m east and measures the bearing of the boat as 340°.
Is the boat closer to the shore than it should be?

11.13 Trigonometry and isosceles triangles

Homework 110 ⌨

1 Find the side or angle marked x.

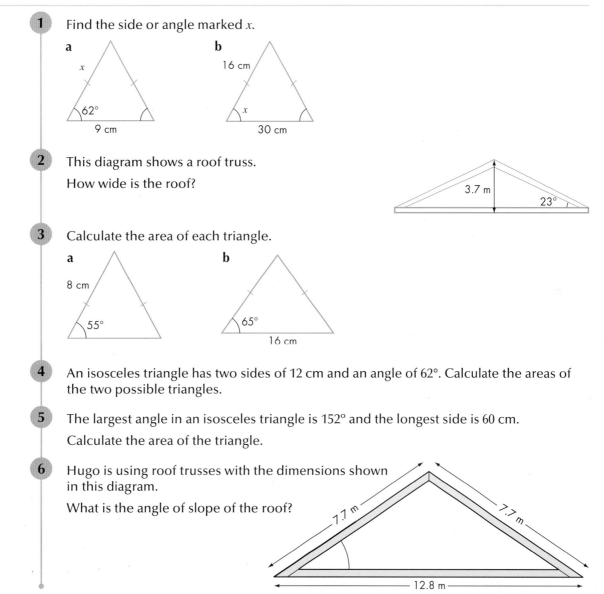

a

x

62°

9 cm

b

16 cm

x

30 cm

2 This diagram shows a roof truss.
How wide is the roof?

3.7 m

23°

3 Calculate the area of each triangle.

a

8 cm

55°

b

65°

16 cm

4 An isosceles triangle has two sides of 12 cm and an angle of 62°. Calculate the areas of the two possible triangles.

5 The largest angle in an isosceles triangle is 152° and the longest side is 60 cm.
Calculate the area of the triangle.

6 Hugo is using roof trusses with the dimensions shown in this diagram.
What is the angle of slope of the roof?

7.7 m

7.7 m

12.8 m

12 Geometry and measures: Similarity

12.1 Similar triangles

Homework 12A

1 These diagrams are drawn to scale. What is the scale factor of the enlargement in each case?

a

b

2 These triangles are similar.

a Show that these triangles are similar.

b Give the ratio of the sides.

c Which angle corresponds to angle C?

d Which side corresponds to side QP?

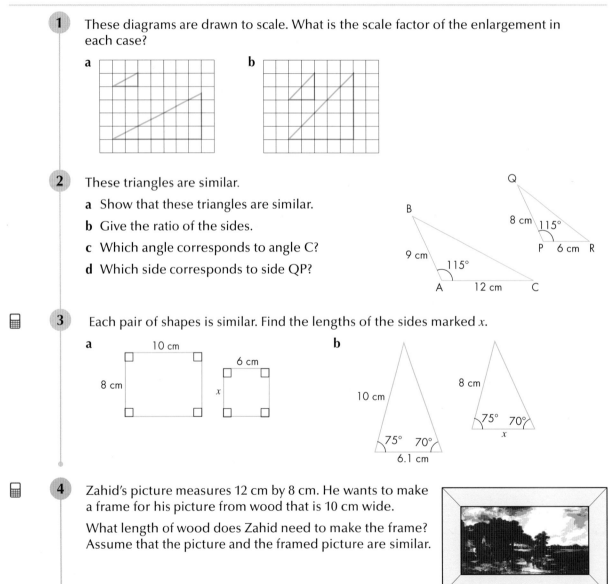

3 Each pair of shapes is similar. Find the lengths of the sides marked x.

a

10 cm

6 cm

8 cm

x

b

10 cm

8 cm

75° 70°

6.1 cm

75° 70°

x

4 Zahid's picture measures 12 cm by 8 cm. He wants to make a frame for his picture from wood that is 10 cm wide.

What length of wood does Zahid need to make the frame? Assume that the picture and the framed picture are similar.

5 Triangle ABC is similar to triangle CDE.
The length of BD is 25 cm.
Work out the lengths of BC and CD.

6 Triangle ABC is similar to triangle DBE
Work out the length of AC.

7 **a** Show that these two triangles are similar.
b What is the ratio of their sides?
c Use Pythagoras' theorem to calculate the length of side PR of triangle PQR.
d Write down the length of the side WY of triangle WXY.

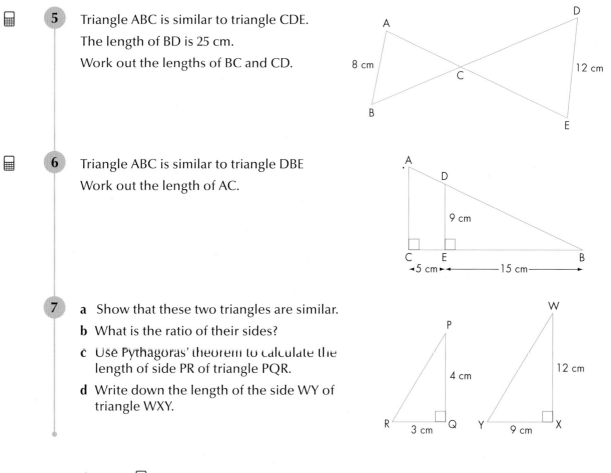

Homework 12B

1 In each of the cases below, state a pair of similar triangles and calculate the length marked x. Separate the similar triangles if it makes it easier for you.

a

b

2 Calculate the lengths of the sides marked x and y in these diagrams.

a

b

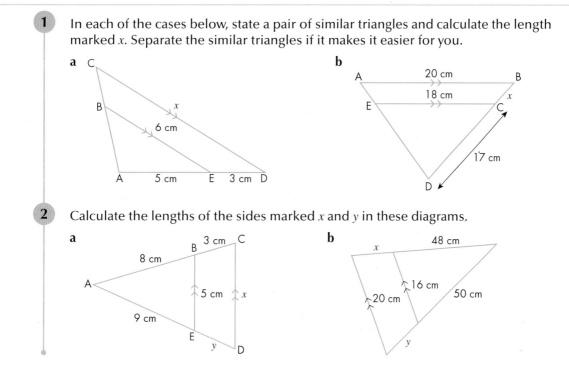

3 A man of height 158 cm stands next to a lamppost and notices that his shadow is 90 cm. At the same time, the lamppost casts a shadow of 2.1 m. Calculate the height of the lamppost.

4 Jamie is making this metal frame for a garden slide.

In the diagram, triangles ABC is similar to triangle ADE.

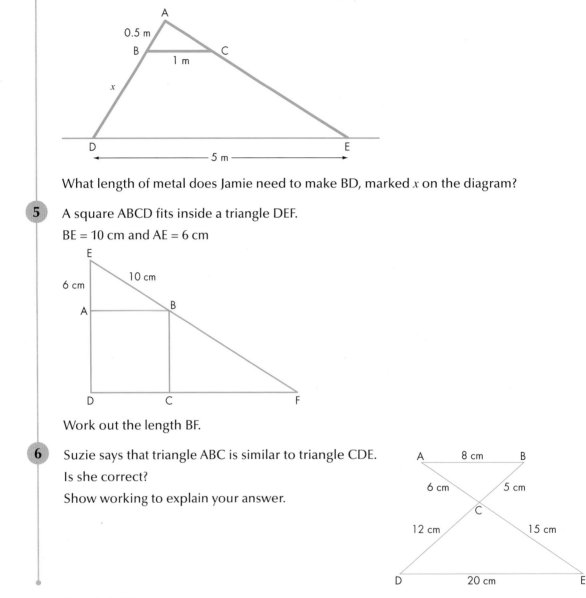

What length of metal does Jamie need to make BD, marked x on the diagram?

5 A square ABCD fits inside a triangle DEF.

BE = 10 cm and AE = 6 cm

Work out the length BF.

6 Suzie says that triangle ABC is similar to triangle CDE.

Is she correct?

Show working to explain your answer.

Homework 12C 🖩

1 Calculate the lengths marked x in the diagrams below.

a   **b**   **c**

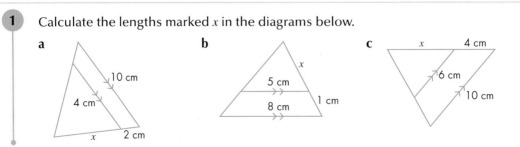

2 Calculate the lengths marked x and y in these diagrams.

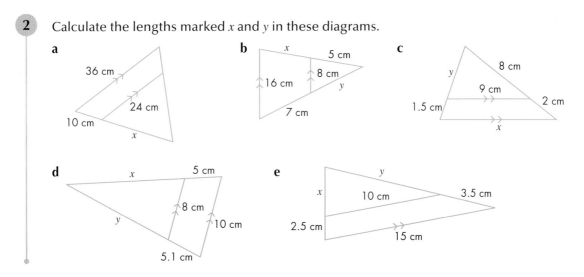

a 36 cm, 24 cm, 10 cm, x

b x, 5 cm, 8 cm, 16 cm, y, 7 cm

c y, 8 cm, 9 cm, 1.5 cm, 2 cm, x

d x, 5 cm, 8 cm, y, 10 cm, 5.1 cm

e y, x, 10 cm, 3.5 cm, 2.5 cm, 15 cm

12.2 Areas and volumes of similar shapes

Homework 12D 🖩

1 The length ratio between two similar solids is 3 : 7.

    a What is the area ratio between the solids?

    b What is the volume ratio between the solids?

2 Copy and complete this table.

Linear scale factor	Linear ratio	Linear fraction	Area scale factor	Volume scale factor
4	1 : 4	$\frac{4}{1}$		
$\frac{1}{2}$				
	10 : 1			
			36	
				125

3 Don marks out a flower bed with an area of 20 cm². He then decides that he wants to use more space. What would be the area of a similar flower bed with lengths that are four times the corresponding lengths of the first shape?

4 A brick has a volume of 400 cm³. What would be the volume of a similar brick with lengths that are:

    a three times the corresponding lengths of the first brick

    b five times the corresponding lengths of the first brick?

5 A tin of paint, 12 cm high, holds 4 litres of paint. Show that a similar tin 36 cm high would hold 108 litres of paint.

6 A model statue is 15 cm high and has a volume of 450 cm³. The real statue is 4.5 m high. What is the volume of the real statue? Give your answer in m³.

7 Tim has a large tin full of paint and wants to transfer the paint into a number of smaller tins, like those in the diagrams.

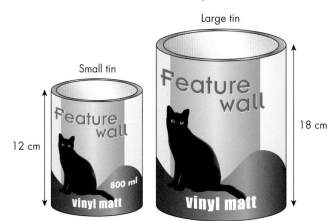

How many small tins can he fill from one large tin?

8 The side length of a cube increases by 10%.

a What is the percentage increase in the total surface area of the cube?

b What is the percentage increase in the volume of the cube?

9 Standard and large gift boxes are similar.

The length of a large gift box is 15 cm.

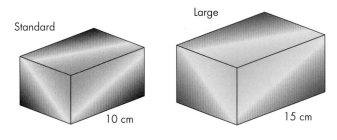

The volume of the standard box is 240 cm³.

Which of the following is the correct volume of the large box?

a 360 cm³   **b** 540 cm³   **c** 720 cm³   **d** 810 cm³

Homework 12E

1 A firm makes three sizes of similar-shaped bottles. Their volumes are:

Small: 330 cm³   Medium: 1000 cm³   Large: 2000 cm³.

a The medium bottle is 20 cm high. Calculate the heights of the other two bottles.

b The firm designs a label for the large bottle and wants the labels on the other two bottles to be similar. If the area of the label on the large bottle is 100 cm², work out the area of the labels on the other two bottles.

2 It takes 1 kg of grass seed to seed a lawn that is 20 m long. How much seed will be needed for a similarly shaped lawn that is 10 m long?

3 The mast on Erin's scale model yacht is 40 cm high. She dreams of owning the real yacht which has a mast that is 4 m high.

  a The sail on her model yacht has area 600 cm^2. What would be the area of the sail on the real yacht? Give your answer in m^2.

  b The volume of the hull on the real yacht is 20 m^3. What is volume of the hull on her model yacht? Give your answer in cm^3.

4 The ratio of the height of P to the height of Q is 5 : 4.
The volume of P is 150 cm^3. Calculate the volume of Q.

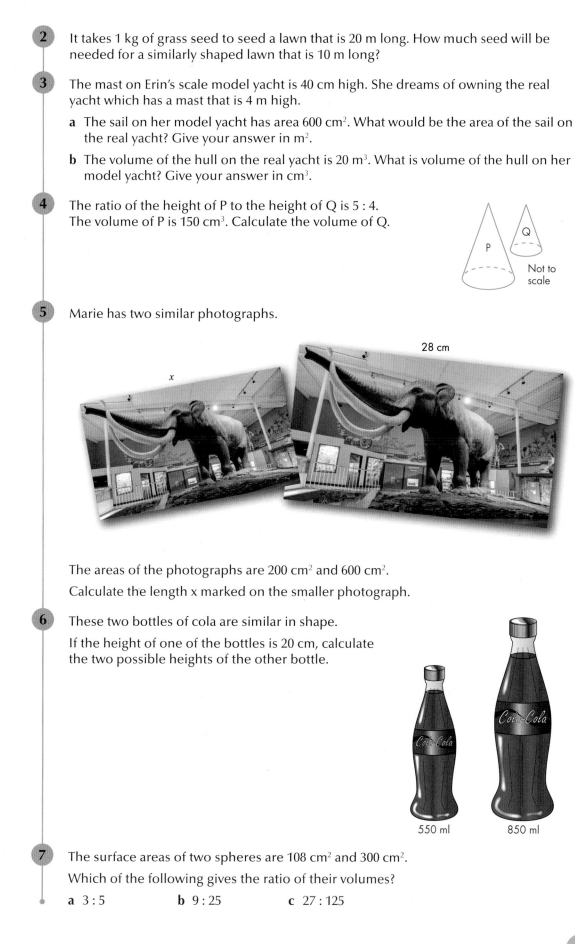

Not to scale

5 Marie has two similar photographs.

28 cm

x

The areas of the photographs are 200 cm^2 and 600 cm^2.

Calculate the length x marked on the smaller photograph.

6 These two bottles of cola are similar in shape.

If the height of one of the bottles is 20 cm, calculate the two possible heights of the other bottle.

550 ml   850 ml

7 The surface areas of two spheres are 108 cm^2 and 300 cm^2.

Which of the following gives the ratio of their volumes?

  a 3 : 5   **b** 9 : 25   **c** 27 : 125

13 Probability: Exploring and applying probability

13.1 Experimental probability

Homework 13A

1 Kate picks a ball at random from a bag that contains six red and four white balls. The table shows her results.

Number of picks	10	20	50	100	500
Number of white balls	2	6	18	42	192

a Calculate the experimental probability of picking a white ball at each stage that Kate recorded her results.

b What is the theoretical probability of picking a white ball from the bag?

c If Kate picked a ball out of the bag a total of 5000 times, how many white balls would you expect her to pick?

2 William made a six-sided spinner. He tested it by spinning it 600 times. The table shows the results.

Number of spinner lands on	1	2	3	4	5	6
Number of times	98	152	85	102	62	101

a Work out the relative frequency of the spinner landing on each number.

b How many times would you expect each number to occur if the spinner is fair?

c Do you think that the spinner is fair? Give a reason for your answer.

3 A sampling bottle is a sealed bottle with a clear plastic tube at one end. When the bottle is tipped up, one of the balls inside wall fall into the tube. Evie's sampling bottle contains 50 balls which are either red, white or blue.

Evie conducts an experiment to see how many balls of each colour are in the bottle. She keeps a tally of each colour as she sees it and records in the table the totals after 100, 250, 400 and 500 trials.

a Calculate the relative frequencies of each colour for each stage Evie recorded her results. Give the frequencies to 3 sf.

b How many of each colour do you think are in the bottle? Give reasons for your answer.

Red	White	Blue	Total
31	52	17	100
68	120	62	250
102	203	95	400
127	252	121	500

4 Which method, **A, B or C,** would you use to estimate or state the probabilities of **a** to **g**?

A: Equally likely outcomes

B: Survey or experiment

C: Look at historical data

a There will be an earthquake in Japan.

b The next person to walk through the door will be female.

c A Premier League team will win the FA Cup.

d You will win a raffle.

e The next car to drive down the road will be foreign.

f You will have a Maths lesson this week.

g A person picked at random from your school will go abroad for their holiday.

5 A bag contains red, blue and yellow counters. Each counter is numbered either 1 or 2. The table shows the probability of selecting these counters at random from the bag.

	Colour of counter		
Number of counter	Red	Blue	Yellow
1	0.2	0.3	0.1
2	0.2	0.1	0.1

a A counter is taken from the bag at random. What is the probability that:

   i it is red *and* numbered 2   ii it is blue *or* numbered 2

   iii it is red *or* numbered 2?

b There are two yellow counters in the bag. How many counters are in the bag altogether?

6 A survey was carried out for one week on all Route 79 buses to find out how many of the passengers were pensioners.

	Mon	Tue	Wed	Thu	Fri
Passengers	950	730	1255	796	980
Pensioners	138	121	168	112	143

For each day, calculate the probability that the 400th passenger to board the route was a pensioner?

7 Joseph made a six-sided spinner. He tested it out to see if it was fair.

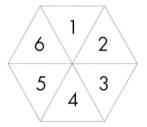

He spun the spinner 240 times and recorded the results in a table.

Number spinner lands on	1	2	3	4	5	6
Frequency	43	38	32	41	42	44

Do you think the spinner is fair? Give reasons for your answer.

8 Aleena tossed a coin 50 times.

She said: "If this is a fair coin, I should get exactly 25 tails." Explain why she is wrong.

13.2 Mutually exclusive and exhaustive outcomes

Homework 13B

1 Say which of these pairs of outcomes are mutually exclusive for a single event.

  a Tossing two heads with two coins *and* tossing two tails with two coins

  b Throwing an even number with a dice *and* throwing an odd number with a dice

  c Drawing a Queen from a pack of cards *and* drawing an Ace from a pack of cards

  d Drawing a Queen from a pack of cards *and* drawing a red card from a pack of cards

  e Drawing a red card from a pack of cards *and* drawing a Heart from a pack of cards

2 Which of the pairs of mutually exclusive events in question **1** are also exhaustive?

3 A letter is to be chosen at random from this set of letter-cards.

M   I   S   S   I   S   S   I   P   P   I

  a What is the probability that the letter chosen is:

    i an S   **ii** a P   **iii** a vowel?

  b Which of these pairs of events are mutually exclusive?

    i Picking an S *and* picking a P   **ii** Picking an S *and* picking an I

    iii Picking an I *and* picking a consonant

  c Which pair of mutually exclusive events in part **b** is also exhaustive?

4 Two of these six people are to be chosen for a job.

    Ann   Joan   Jack   John   Arthur   Ethel

  a List all 15 possible pairs.

  b What is the probability that the pair of people chosen will:

    i both be female   **ii** both be male   **iii** both have the same initial

    iv have different initials?

  c Which of these pairs of outcomes are mutually exclusive?

    i Picking two women *and* picking two men

    ii Picking two people of the same gender *and* picking two people of different genders

    iii Picking two people with the same initial *and* picking two men

    iv Picking two people with the same initial *and* picking two women

  d Which pair of mutually exclusive events in part **c** is also exhaustive?

5 For breakfast I have a choice of toast, porridge or cereal. The probability that I choose toast is $\frac{1}{3}$ and the probability that I choose porridge is $\frac{1}{2}$. What is the probability that I choose cereal?

6 A person is chosen at random. Here is a list of outcomes.

A: The person chosen is male   B: The person chosen is female

C: The person chosen is over **18**   D: The person chosen is under **16**

E: The person chosen has a degree   F: The person chosen is a teacher

State whether each pair of outcomes is:

a mutually exclusive   **b** exhaustive.

If they are not mutually exclusive, give reasons why.

    i A and B   **ii** A and C   **iii** B and D   **iv** C and D

    v D and F   **vi** E and F   **vii** E and D   **viii** A and E

    ix C and F   **x** C and E

7 An amateur weather man, Steve, records the weather in his village for one year.

He knows that the probability of a windy day is 0.4 and that the probability of a rainy day is 0.6.

Steve says: "This means it will be either rainy or windy each day as 0.4 + 0.6 = 1, which is certain." Show that Steve is wrong.

8 Four brothers, David, Malcolm, Brian and Kevin, regularly run races against each other in the park.

The chance of:

    David winning the race is 0.3.

    Malcolm winning the race is $\frac{1}{5}$.

    Brian winning the race is 45%.

    What is the chance of Kevin winning the race?

9 Gareth always walks, goes by bus or is given a lift by his dad to football training.

If he walks, the probability that he is late for training is 0.4.

If he goes by bus, the probability that he is late for training is 0.5.

Show that it is not necessarily true that if his dad gives him a lift, the probability he is late for training is 0.1.

13.3 Expectation

Homework 13C

1 I roll an ordinary dice 600 times. How many times can I expect to score a 1?

2 I throw a coin 500 times. How many times can I expect to score a tail?

3 I draw a card from a pack of cards and replace it. I do this 104 times. How many times would I expect to get:

    a a red card   **b** a Queen   **c** a red seven   **d** the Jack of Diamonds?

4 The ball in a roulette wheel can land in 37 spaces, which are numbered form 0 to 36 inclusive.

I always bet on the same block of numbers, 1–6.

If I play all evening and there are a total of 111 spins of the wheel in that time, how many times could I expect to win?

5 In a bag there are 20 balls, 10 of which are red, 3 yellow and 7 blue.

I take out a ball at random, note its colour and then replace it. I do this 200 times.

How many times would I expect to get:

a a red ball **b** a yellow or blue ball

c a ball that is not blue **d** a green ball?

6 A sampling bottle contains black and white balls. Lucas knows that the probability of getting a black ball is 0.4.

He takes 200 samples. How many of them would he expect to give a white ball?

7 I have five tickets for a raffle. The probability that I win the main prize is 0.003.

How many raffle tickets were sold altogether?

8 **a** Fred is about to take his driving test. The chance that he passes is $\frac{1}{3}$. His sister says:

"You are sure to pass within three attempts because $3 \times \frac{1}{3} = 1$." Explain why his sister is wrong.

b If Fred does fail, would you expect the chance that he passes next time to increase or decrease? Give reasons for your answer.

9 Kara rolls two dice 200 times.

a How many times would she expect to roll a double?

b How many times would she expect to roll a total score greater than 7?

10 An opinion poll uses a sample of 200 voters in one area. 112 of them said they would vote for Party A.

a There are a total of 50 000 voters in the area. If they all voted, how many would you expect to vote for Party A?

b The poll is accurate within 10%. Can Party A be confident of winning?

11 The ball in a roulette wheel can land in 37 spaces, which are numbered form 0 to 36 inclusive. I always bet on a prime number.

If I play the game 100 times, how many times could I expect to win?

12 A headteacher is told that the probability of any student being left-handed is 0.14.

Show how she can work out how many of her students she should expect to be left-handed.

13.4 Probability and two-way tables

Homework 13D ▤

1. This two-way table shows the number of children and the number of computers in 40 homes in the same road.

		Number of children			
		0	1	2	3
Number of computers	0	3	0	0	0
	1	4	10	2	2
	2	0	4	9	3
	3	0	0	2	1

   a How many homes have exactly two children and two computers?

   b How many homes altogether have two computers?

   c What percentage of the homes have two computers?

   d What percentage of the homes with just one child have one computer?

2. This two-way table shows the weekly earnings of a set of students during the summer break.

		Male	Female
Earnings per week	£0 ⩽ E < £50	4	1
	£50 ⩽ E < £100	4	1
	£100 ⩽ E < £150	11	4
	£150 ⩽ E < £200	24	14
	£200 ⩽ E < £250	16	10
	E ⩾ £250	2	1

   a What percentage of the male students earned between £100 and £150 per week?

   b What percentage of the female students earned between £100 and £150 per week?

   c Estimate the mean earnings of this set of students.

   d Which group of students has the higher estimated mean earnings: male or female? Explain how you could find this without doing the actual calculation.

3. Elena spins two fair spinners and adds together the numbers scored.

   a Draw a two-way diagram showing all the possible totals.

   b What is the most unlikely score?

   c What is the probability of Elena getting a total of 10?

   d What is the probability of her getting a total of 9 or more?

   e What is the probability of her getting a total that is an even number?

4 Hassan has two hexagonal spinners.

Spinner X is numbered 2, 4, 6, 8, 9 and 11.

Spinner Y is numbered 3, 4, 5, 6, 7, and 8.

What is the probability that, when Hassan spins the two spinners, the product of the two numbers he scores will be greater than 30?

5 Connie planted some tomato plants and kept them in the kitchen. Her husband, Harold, planted some tomato plants in the garden.

The table shows some information about their crops.

	Connie	Harold
Mean Diameter	1.9 cm	4.3 cm
Mean number of tomatoes per plant	23.2	12.3

Use the data to show who had the better crop of tomatoes.

13.5 Probability and Venn diagrams

Homework 13E 🖩

1 P(A) = 0.2 and P(B) = 0.6. Write down:

a P(A′)   **b** P(B′).

2 P(A) = 0.35 and P(B) = 0.45. Write down:

a P(A′)   **b** P(B′).

3 ξ = {1, 2, 3, 4, 5, 6, 7, 8, 9, 10}   A = {1, 3, 5, 7}   B = {1, 3, 4, 8, 9}

a Show this information in a Venn diagram.

b Use your Venn diagram to work out:

   i P(A)   **ii** P(A′)   **iii** P(B)

   iv P(B′)   **v** P(A∪B)   **vi** P(A∩B).

4 In a survey, Katie asked 100 people if they liked crisps (C) and doughnuts (D).

The results are shown in the Venn diagram (ξ).

A person is chosen at random.

a Work out:

   i P(C)   **ii** P(C′)   **iii** P(D)   **iv** P(D′)   **v** P(C∪D)   **vi** P(C∩D).

b Work out the probability that a person likes doughnuts but does not like crisps.

5 The Venn diagram (ξ) shows some probabilities.

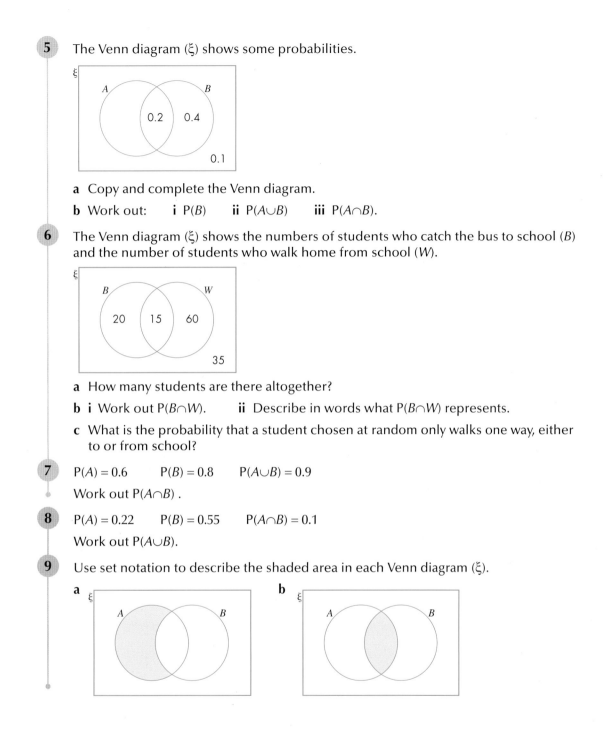

a Copy and complete the Venn diagram.

b Work out: i P(B)   ii P(A∪B)   iii P(A∩B).

6 The Venn diagram (ξ) shows the numbers of students who catch the bus to school (B) and the number of students who walk home from school (W).

a How many students are there altogether?

b i Work out P(B∩W).   ii Describe in words what P(B∩W) represents.

c What is the probability that a student chosen at random only walks one way, either to or from school?

7 $P(A) = 0.6$ $P(B) = 0.8$ $P(A∪B) = 0.9$

Work out $P(A∩B)$.

8 $P(A) = 0.22$ $P(B) = 0.55$ $P(A∩B) = 0.1$

Work out $P(A∪B)$.

9 Use set notation to describe the shaded area in each Venn diagram (ξ).

a

b

14 Number: Powers and standard form

14.1 Powers (indices)

Homework 14A

1 Write these expressions in index notation. Do not work them out yet.

  a $5 \times 5 \times 5 \times 5$ **b** $7 \times 7 \times 7 \times 7 \times 7$ **c** $19 \times 19 \times 19$

  d $4 \times 4 \times 4 \times 4 \times 4$ **e** $1 \times 1 \times 1 \times 1 \times 1 \times 1 \times 1$ **f** $8 \times 8 \times 8 \times 8 \times 8$

  g 6 **h** $11 \times 11 \times 11 \times 11 \times 11 \times 11$

  i $0.9 \times 0.9 \times 0.9 \times 0.9$ **j** $999 \times 999 \times 999$

2 Write out each of these power terms in full. Do not work them out yet.

  a 4^5 **b** 8^4 **c** 5^3 **d** 9^6 **e** 1^{11}

  f 7^3 **g** 5.2^3 **h** 7.5^3 **i** 7.7^4 **j** $10\,000^3$

3 Use the power key on your calculator (or any method you prefer), to work out the value of each power term in question **1**.

4 Use the power key on your calculator (or any method you prefer), to work out the value of each power term in question **2**.

5 A storage container is in the shape of a cube.

  The length of the container is 1 m.

  a Work out the total storage space in the container. Use the formula for the volume of a cube. Volume = (length of edge)^3

  b The box is to be used for storing dolls. The volume taken up by a doll is 0.03 m^3.

    Work out the volume of free storage space when 24 dolls are packed into the container.

6 Write each number as a power of a different number. The first one has been done for you.

  a $27 = 3^3$ **b** 16 **c** 125 **d** 64

7 Work out the values of each power term. Do not use a calculator.

  a 7^0 **b** 9^1 **c** 17^0 **d** 1^{91} **e** 10^5

8 Using your calculator, or otherwise, work out the values of each power term.

  a $(-2)^3$ **b** $(-1)^{11}$ **c** $(-3)^4$ **d** $(-5)^3$ **e** $(-10)^6$

9 $5^5 = 3125$ $5^6 = 15\,625$ $5^7 = 78\,125$ $5^8 = 390\,625$

  Write down the last three digits of these powers of 5.

  a 5^{99} **b** 5^{100}

14.2 Rules for multiplying and dividing powers

Homework 14B

1 Write each of these as a single power of 7.

  a $7^3 \times 7^2$ **b** $7^3 \times 7^6$ **c** $7^4 \times 7^3$ **d** 7×7^5 **e** $7^5 \times 7^9$ **f** 7×7^7

2 Write each of these as a single power of 5.

  a $5^6 \div 5^2$ **b** $5^8 \div 5^2$ **c** $5^4 \div 5^3$ **d** $5^5 \div 5^5$ **e** $5^6 \div 5^4$

3 Simplify these and write each of them as a single power of a.

  a $a^2 \times a$ **b** $a^3 \times a^2$ **c** $a^4 \times a^3$ **d** $a^6 \div a^2$ **e** $a^3 \div a$ **f** $a^5 \div a^4$

4 Write down a possible pair of values for x and y:

  a $a^x \times a^y = a^6$ **b** $a^x \div a^y = a^6$

5 Simplify each expression.

  a $3a^4 \times 5a^2$ **b** $3a^4 \times 7a$ **c** $5a^4 \times 6a^2$ **d** $3a^2 \times 4a^7$ **e** $5a^4 \times -5a^2 \times 5a^2$

6 Simplify each expression.

  a $8a^5 \div 2a^2$ **b** $12a^7 \div 4a^2$ **c** $25a^6 \div 5a$ **d** $48a^8 \div 6a^{-1}$ **e** $24a^6 \div 8a^{-2}$ **f** $36a \div 6a^5$

7 Simplify each expression.

  a $3a^3b^2 \times 4a^3b$ **b** $7a^3b^5 \times 2ab^3$ **c** $4a^3b^5 \times 5a^4b^{-1}$

  d $12a^3b^5 \div 4ab$ **e** $24a^3b^5 \div 6a^2b^{-3}$

8 Simplify these expressions.

  a $\dfrac{8a^6b^5}{4a^3b^4}$ **b** $\dfrac{3a^3b^2c^4 \times 4a^2bc^6}{6a^4b^4c^8}$ **c** $\dfrac{2a^3b^4c \times 3a^3b^2c \times 9c^3}{6a^2bc}$

9 Write down *two* possible:

  a multiplication questions with an answer of $18x^3y^4$

  b division questions with an answer of $18x^3y^4$.

10 a and b are different prime numbers. What is the smallest value of a^2b^2?

11 Use the general rule for dividing powers of the same number, $a^x \div a^y = a^{x-y}$,

  to prove that any number raised to the power -1 is the reciprocal of that number.

14.3 Standard form

Homework 14C

1 Write down the answers without using a calculator.

   a 400×300 **b** 50×4000 **c** 70×200 **d** 30×700

   e $(30)^2$ **f** $(50)^3$ **g** $(200)^2$ **h** 40×150

   i 60×5000 **j** 30×250 **k** 700×200

2 Write down the answers without using a calculator.

   a $4000 \div 800$ **b** $9000 \div 30$ **c** $7000 \div 200$ **d** $8000 \div 200$

   e $2100 \div 700$ **f** $9000 \div 60$ **g** $700 \div 50$ **h** $3500 \div 70$

   i $3000 \div 500$ **j** $30\,000 \div 2000$ **k** $5600 \div 1400$ **l** $6000 \div 30$

3 Write down the value of each expression.

   a 2.3×10 **b** 2.3×100 **c** 2.3×1000 **d** $2.3 \times 10\,000$

4 Write down the value of each expression.

   a 5.4×10 **b** 5.4×10^2 **c** 5.4×10^3 **d** 5.4×10^4

5 Write down the value of each expression.

   a $2.3 \div 10$ **b** $2.3 \div 100$ **c** $2.3 \div 1000$ **d** $2.3 \div 10\,000$

6 Write down the value of each expression.

   a $5.4 \div 10$ **b** $5.4 \div 10^2$ **c** $5.4 \div 10^3$ **d** $5.4 \div 10^4$

7 Evaluate each expression.

   a 3.5×100 **b** 2.15×10 **c** 6.74×1000 **d** 4.63×10

   e 30.145×10 **f** 78.56×1000 **g** 6.42×10^2 **h** 0.067×10

   i 0.085×10^3 **j** 0.798×10^5 **k** 0.658×1000 **l** 215.3×10^2

   m 0.889×10^6 **n** 352.147×10^2 **o** 37.2841×10^3 **p** 34.28×10^6

8 Evaluate each expression.

   a $4538 \div 100$ **b** $435 \div 10$ **c** $76\,459 \div 1000$ **d** $643.7 \div 10$

   e $4228.7 \div 100$ **f** $278.4 \div 1000$ **g** $246.5 \div 10^2$ **h** $76.3 \div 10$

   i $76 \div 10^3$ **j** $897 \div 10^5$ **k** $86.5 \div 1000$ **l** $1.5 \div 10^2$

   m $0.8799 \div 10^6$ **n** $23.4 \div 10^2$ **o** $7654 \div 10^3$ **p** $73.2 \div 10^6$

9 Evaluate each expression.

   a 7.3×10^2 **b** 3.29×10^5 **c** 7.94×10^3 **d** 6.8×10^7

   e $3.46 \div 10^2$ **f** $5.07 \div 10^4$ **g** $2.3 \div 10^4$ **h** $0.89 \div 10^3$

10 The diameter of Venus is approximately 7.5×10^3 miles.

   The diameter of Saturn is approximately 7.5×10^4 miles.

   The diameter of Earth is approximately 7.9×10^3 miles.

   Without working out the answers, explain how you can tell which of these planets is the biggest.

11 A number is between 10 000 and 100 000. It is written in the form 2.5×10^n.

What is the value of n?

Homework 14D

1 Write down the value of each expression.

    a 2.3×0.1 **b** 2.3×0.01 **c** 2.3×0.001 **d** 2.3×0.0001

2 Write down the value of each expression.

    a 5.4×10^{-1} **b** 5.4×10^{-2} **c** 5.4×10^{-3} **d** 5.4×10^{-4}

3 Work out the value of each expression.

    a $2.3 \div 0.1$ **b** $2.3 \div 0.01$ **c** $2.3 \div 0.001$ **d** $2.3 \div 0.0001$

4 Work out the value of each expression.

    a $5.4 \div 10^{-1}$ **b** $5.4 \div 10^{-2}$ **c** $5.4 \div 10^{-3}$ **d** $5.4 \div 10^{-4}$

5 Write these numbers out in full.

    a 3.5×10^2 **b** 4.15×10 **c** 5.7×10^{-3} **d** 3.89×10^{-2}

    e 4.6×10^3 **f** 8.6×10 **g** 3.97×10^5 **h** 3.65×10^{-3}

6 Write these numbers in standard form.

    a 780   **b** 0.435   **c** 67 800   **d** 7 400 000 000

    e 30 780 000 000   **f** 0.000 427 8   **g** 6450   **h** 0.047

    i 0.000 12   **j** 96.43   **k** 74.78   **l** 0.004 157 8

7 Write the appropriate numbers in each statement in standard form.

    a Last year there were 24 673 000 vehicles licensed in the UK.

    b Daryl John was one of 15 282 runners to complete the Boston Marathon.

    c Last year 613 000 000 000 passenger kilometres were completed on British roads.

    d The Sun is 93 million miles away from Earth. The next closest star to the Earth is Proxima Centuri at a distance of about 24 million million miles.

    e A scientist claims to be working with a new particle weighing only 0.000 000 000 000 65 g.

8 How many times smaller is 1.7×10^2 than 1.7×10^5?

9 How many times greater is 9.6×10^7 than 4.8×10^3?

10 How many times greater is 1.2×10^5 than 3000?

11 The speed of light is approximately 3.00×10^8 m/s.

It takes about 1.3 seconds for light to travel to the moon.

Work out the approximate distance to the moon in kilometres.

1. These numbers are not in standard form. Write them in standard form.

   a 36.8×10^3

   b 0.09×10^3

   c 41.7×10^{-2}

   d 0.08×10^{-3}

   e 24×10

   f $3 \times 4 \times 10^4$

   g $3 \times 10^3 \times 25$

   h 170×10^{-3}

   i 48 million

2. Work these out. Give your answers in standard form.

   a $3.1 \times 10^2 \times 5.1 \times 10^3$

   b $2.4 \times 10^5 \times 3.2 \times 10^3$

   c $9.6 \times 10^5 \times 7.6 \times 10^3$

   d $6.3 \times 10^{-4} \times 3.4 \times 10^2$

   e $(2.1 \times 10^5)^2$

   f $(7.8 \times 10^{-3})^2$

   g $5.6 \times 10 \times 2.7 \times 10$

   h $2.4 \times 10^{-5} \times 2.6 \times 10^8$

   i $7.3 \times 10^4 \times 2.7 \times 10^{-3}$

3. Work these out. Give your answers in standard form.

   a $(9 \times 10^8) \div (3 \times 10^4)$

   b $(2.7 \times 10^7) \div (9 \times 10^3)$

   c $(5.5 \times 10^4) \div (1.1 \times 10^{-2})$

   d $(4.2 \times 10^{-9}) \div (3 \times 10^{-8})$

4. Work out the value of each expression, giving your answers in standard form.

   a $\dfrac{8 \times 10^9}{4 \times 10^7}$

   b $\dfrac{12 \times 10^6}{3 \times 10^4}$

   c $\dfrac{2.8 \times 10^7}{7 \times 10^{-4}}$

5. The population of Africa in 2013 was approximately 1×10^9.

   It is expected to reach 1.8×10^9 in 2050.

   By how much is the population of Africa expected to rise between 2013 and 2050?

   Give your answer in millions.

6. A number, when written in standard form, is greater than 1 million and less than 5 million.

   Write down a possible value of the number in standard form.

7. These four numbers are written in standard form.

   3.5×10^5 1.2×10^3 7.3×10^2 4.8×10^4

   a Work out the largest possible answer when two of these numbers are multiplied together.

   b Work out the smallest possible answer when two of these numbers are added together.

   Give your answers in standard form.

8. The Moon is a sphere with a radius of 1.08×10^3 miles.

   Using the formula for the surface area of a sphere:

   $$\text{volume} = \frac{4}{3}\pi r^3$$

   calculate the volume of the Moon.

   Give your answer to an appropriate degree of accuracy.

15 Algebra: Equations and inequalities

15.1 Linear equations

Homework 15A

1 Solve these equations.

a $\frac{g}{3} + 2 = 8$ **b** $\frac{m}{4} - 5 = 2$ **c** $\frac{h}{8} - 3 = 5$ **d** $\frac{2h}{3} + 3 = 7$ **e** $\frac{3t}{4} - 3 = 6$

f $\frac{3x}{4} - 1 = 8$ **g** $\frac{x+5}{3} = 2$ **h** $\frac{t+12}{2} = 5$ **i** $\frac{w-3}{5} = 3$ **j** $\frac{y-9}{2} = 3$

2 The solution to the equation $\frac{3x}{4} + 3 = 9$ is $x = 8$.

Make up two more different equations of the form $\frac{ax}{b} \pm c = d$ for which x is also 8, where a, b, c and d are positive whole numbers.

3 Solve these equations.

a $\frac{2x-1}{3} = 5$ **b** $\frac{5t-4}{2} = 3$ **c** $\frac{4m+1}{5} = 5$ **d** $\frac{8p-6}{5} = 2$

e $\frac{5x+1}{4} = 4$ **f** $\frac{17+2t}{9} = 1$ **g** $\frac{2+4x}{3} = 4$ **h** $\frac{8-2x}{11} = 1$

4 Eight friends went for a meal in a restaurant. The bill was £x but, due to some service problems, they received a reduction of £2 per person.
They shared the final bill equally between them. Each person paid £11.25.
a Set this problem up as an equation.
b Solve the equation and find the cost of the bill before the reduction.

5 Mr O'Leary asked his class to solve the equation $\frac{5x+3}{6} = 8$.

Beth wrote:
$$5x + 3 = 6 \times 8$$
$$5x + 3 - 3 = 48 - 3$$
$$5x = 45$$
$$5x \div 5 = 45 \div 5$$
$$x = 9$$

Arabella wrote:
$$\frac{5x}{6} = 8 - 3$$
$$5x = 5 \times 6$$
$$5x = 30$$
$$5x \div 5 = 15 \div 5$$
$$x = 3$$

Who is correct? Give reasons why.

Homework 15B 🖩

1 Solve each of these equations. Give your answers as fractions or decimals as appropriate. Remember to check that each answer works for its original equation. Use your calculator if necessary.

a $3(x+6) = 15$ **b** $6(x-5) = 30$ **c** $5(4x+3) = 45$ **d** $3(4y-7) = 15$

e $3(4t+2) = 18$ **f** $4(6x+5) = 8$ **g** $5(2k+3) = 35$ **h** $5(2x+8) = 30$

i $3(2y-7) = 21$ **j** $3(2t-5) = 27$ **k** $8(2x-7) = 16$ **l** $8(3x-4) = 16$

m $4(x+7) = 8$ **n** $3(x-5) = -24$ **o** $5(t+3) = 15$ **p** $4(3x-13) = 8$

q $5(4t+3) = 20$ **r** $2(5x-3) = -16$ **s** $4(6y-8) = -8$ **t** $3(2x+7) = 9$

2 A rectangular room is 4 m longer than it is wide. The perimeter is 28 m.

The cost of the carpet for this room was £607.50.

What is the cost of the carpet per square metre?

3 Mike has been asked to solve the equation $a(bx + c) = 60$.

He knows that x is an even number and that the values of a, b and c are 2, 4 and 5, but he doesn't know which value is which.

Find the values for a, b and c for this equation.

4 Mrs Chan is writing some equations on the board for her class to solve at the start of the lesson. So far she has written:

$5(2x + 3) = 13$

$2(5x + 3) = 13$

Zak says, "That's easy – both equations have the same solution, $x = 2$."

Is Zak correct? If not, give the correct answers and explain his mistakes.

Homework 15C

1 Solve each of these equations.

a $3x + 4 = x + 6$ **b** $4y + 3 = 2y + 5$ **c** $5a - 2 = 2a + 4$

d $6t + 5 = 2t + 25$ **e** $8p - 3 = 3p + 12$ **f** $5k + 4 = 2k + 13$

g $2(d + 4) = d + 15$ **h** $4(x - 3) = 3(x + 3)$ **i** $2(3y + 2) = 5(2y - 4)$

2 Olivia wants to find the mass of some unlabelled cans of rice pudding.

She can't find her set of weights, but she has some balance scales and some other labelled containers.

A jar of jam has a mass of 454 g and a can of beans has a mass of 120 g.

By using trial and improvement, she finds that five cans of rice pudding and one can of beans balance with three cans of rice pudding and two jars of jam.

What is the mass of one can of rice pudding?

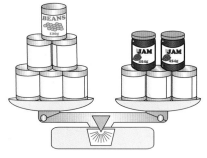

> **Hints and tips** For questions **2** to **4**, set up equations, make them equal and solve.

3 Find the perimeter of this isosceles triangle.

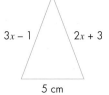

$3x - 1$ $2x + 3$

5 cm

4 John says: "I am thinking of a number. I multiply it by 3 and subtract 6."

Zubert says: "I am thinking of a number. I multiply it by 5 and add 2."

John and Zubert find that they both thought of the same number and both got the same final answer.

What number did they think of?

5 Solve each of these equations.

a $3(2b - 1) + 25 = 4(3b + 1)$ **b** $3(4c + 1) - 17 = 2(3c + 2) - 3c$

6 Solve the equation $(x - 2)(x + 3) = (x + 5)(x - 3)$.

7 **a** Show that the equation $4(3x + 5) = 3(4x + 2)$ cannot be solved.

b Show that there are an infinite number of solutions to the equation $5(2x + 8) = 2(5x + 20)$.

15.2 Elimination method for simultaneous equations

Homework 15D

1 Solve each pair of simultaneous equations by the elimination method.

a $x + 3y = 11$ **b** $3x + 4y = 25$ **c** $3x - 2y = 8$
   $x + 2y = 9$ $3x + 2y = 17$ $4x + 2y = 20$

2 Solve each pair of simultaneous equations by the elimination method.

a $4a + b = 19$ **b** $9c + d = 58$ **c** $5e - 2f = 29$
   $4a + 5b = 31$ $9c - d = 50$ $e + 2f = 13$

3 Solve this pair of simultaneous equations.

$16x - 14y = 76$

$5x - 14y = -34$

15.3 Substitution method for simultaneous equations

Homework 15E

1 Solve each pair of simultaneous equations by the substitution method.

a $2x + 4y = 24$ **b** $3x + y = 22$ **c** $6x - 3y = 18$
   $y = x + 3$ $y = 5x - 26$ $y = x - 2$

2 Solve each pair of simultaneous equations by the substitution method.

a $3x + 4y = 19$ **b** $6x - 2y = 24$ **c** $2x - y = 10$
   $y = 6 - 2x$ $x = 20 - 5y$ $x = 4y - 2$

3 Solve each pair of simultaneous equations by either elimination or substitution.

a $6x + 10y = 72$ **b** $5x + y = 14$
   $y = 31 - 4x$ $5x - 6y = -14$

15.4 Balancing coefficients to solve simultaneous equations

Homework 15F

1 Solve each pair of simultaneous equations.

   a $3x + 2y = 12$ **b** $4x + 3y = 37$ **c** $2x + 3y = 19$ **d** $5x - 2y = 14$

      $4x - y = 5$ $2x + y = 17$ $6x + 2y = 22$ $3x - y = 9$

2 Solve each pair of simultaneous equations.

   a $6x + 5y = 23$ **b** $3x - 4y = 13$ **c** $8x - 2y = 14$ **d** $5x + 2y = 33$

      $5x + 3y = 18$ $2x + 3y = 20$ $6x + 4y = 27$ $4x + 5y = 23$

15.5 Using simultaneous equations to solve problems

Homework 15G

Read each situation carefully, then write a pair of simultaneous equations that you can use to solve the problem.

1 A book and a CD cost £14.00 together. The CD costs £7 more than the book. How much does each cost?

2 It costs two adults and three children £28.50 to go to the cinema.

It costs three adults and two children £31.50 to go to the cinema.

Let the price of an adult ticket be £x and the price of a child's ticket be £y.

   a Using £x to represent the cost of an adult ticket and £y to represent the cost of a child ticket, set up a pair of simultaneous equations to represent the above information.

   b Solve your equations for x and y.

3 Ina wants to buy some snacks for her friends. She works out from the labels that two cakes and three bags of peanuts contain 63 g of fat and that one cake and four bags of peanuts contain 64 g of fat. How many grams of fat are there in:

   a one cake   **b** one bag of peanuts?

4 Ten second-class and six first-class stamps cost £4.96.

Eight second-class and ten first-class stamps cost £5.84.

How much would I pay for three second-class and four first-class stamps?

5 Two people bought cola and chocolate at the local store. Henri paid £4.37 for six cans of cola and five chocolate bars. Evie paid £2 for three cans of cola and two chocolate bars.

How much would it cost Mark to buy two cans of cola and a chocolate bar?

6 Three bags of sugar and four bags of rice have a mass of 12 kg. Five bags of sugar and two bags of rice have a mass of 13 kg. What is the mass of two bags of sugar and five bags of rice?

7 The difference between my son's age and my age is 28 years.

Five years ago my age was double that of my son.

Let my age now be x and my son's age now be y.

a Explain why $x - 5 = 2(y - 5)$. **b** Find the values of x and y.

8 Four apples and two oranges cost £2.04.

Five apples and one orange costs £1.71.

Marcus buys four apples and eight oranges.

How much change will he get from a £10 note?

9 Five bags of compost and four bags of pebbles have a mass of 340 kg.

Three bags of compost and five bags of pebbles have a mass of 321 kg.

Carol needs six bags of compost and eight bags of pebbles.

Her trailer has a safe working load of 500 kg.

Can Carol transport all the bags safely on her trailer?

10 Here are four equations.

A: $2x - y = 8$

B: $3x + y = 19$

C: $4x + y = 16$

D: $3x + 2y = 7$

Here are four sets of (x, y) values.

(5.4, 2.8)   (4, 0)   (−3, 28)   (5, −4)

Match each pair of (x, y) values to a **pair** of equations.

15.6 Linear inequalities

Homework 15H

1 Solve the following linear inequalities.

a $x + 3 < 8$ **b** $t - 2 > 6$ **c** $p + 3 > 11$ **d** $4x - 5 < 7$

e $3y + 4 < 22$ **f** $\dfrac{x + 3}{2} < 8$ **g** $\dfrac{t - 2}{5} > 7$ **h** $2(x - 3) < 14$

i $4(3x + 2) < 32$ **j** $5(4t - 1) \$ 30$ **k** $3x + 1 > 2x - 5$ **l** $6t - 5 < 4t + 3$

m $2y - 11 < y - 5$ **n** $3x + 2 > x + 3$ **o** $4w - 5 < 2w + 2$ **p** $2(5x - 1) < 2x + 3$

2 Write down the values of x that satisfy each of the following.

a $x - 2 \leqslant 3$, where x is a positive integer

b $x + 3 < 5$, where x is a positive, even integer

c $2x - 14 < 38$, where x is a square number

d $4x - 6 \leqslant 15$, where x is a positive, odd number

e $2x + 3 < 25$, where x is a positive, prime number

3 Frank went into town with £6. He bought three cans of cola and lent his brother £3.

When he reached home he put 50p in his piggy bank.

What was the most a can of cola could cost?

4 Solve the following linear inequalities.

a $9 < 4x + 1 < 13$ **b** $2 < 3x - 1 < 11$ **c** $-3 < 4x + 5 < 21$

d $2 \leqslant 3x - 4 < 15$ **e** $10 \leqslant 2x + 3 < 18$ **f** $-5 \leqslant 4x - 7 < 8$

g $3 \leqslant 5x - 7 \leqslant 13$ **h** $8 \leqslant 2x + 3 < 19$ **i** $7 \leqslant 5x + 3 < 24$

5 The perimeter of this rectangle is greater than 10 cm but less than 16 cm.

What are the limits of the area?

2x – 1

x

6 A teacher asks six students to stand at the front of the class and hold up these inequality cards.

$x > 0$ $x < 2$ $x \geqslant 3$ $x = 2$ $x = 3$ $x < 9$

She writes 'TRUE' on one side of the board and 'FALSE' on the other side.

She asks the other students to call out a number, and the students holding the cards have to stand on the 'TRUE' side if their inequality card is true for the number, or on the 'FALSE' side if it isn't.

a A student calls out '2' and the students with the cards all go to the correct side.

   i Which cards are held by the students on the 'TRUE' side?

   ii Which cards are held by the students on the 'FALSE' side?

b Find a number that would satisfy this grouping.

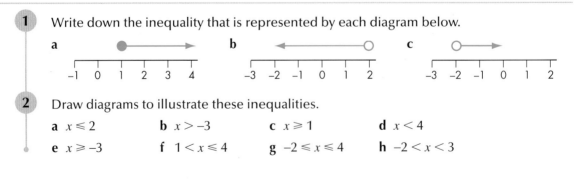

True

$x \geqslant 3$ $x < 9$ $x > 0$

False

$x < 2$ $x = 2$ $x = 3$

Homework 15I

1 Write down the inequality that is represented by each diagram below.

2 Draw diagrams to illustrate these inequalities.

a $x \leqslant 2$ **b** $x > -3$ **c** $x \geqslant 1$ **d** $x < 4$

e $x \geqslant -3$ **f** $1 < x \leqslant 4$ **g** $-2 \leqslant x \leqslant 4$ **h** $-2 < x < 3$

3 Solve the following inequalities and illustrate their solutions on number lines.

   a $x + 5 \geqslant 9$ **b** $x + 4 < 2$ **c** $x - 2 \leqslant 3$ **d** $x - 5 > -2$

   e $4x + 3 \leqslant 9$ **f** $5x - 4 \geqslant 16$ **g** $2x - 1 > 13$ **h** $3x + 6 \leqslant 3$

   i $3(2x + 1) < 15$ **j** $\dfrac{x + 1}{2} \leqslant 2$ **k** $\dfrac{x - 3}{3} > 7$ **l** $\dfrac{x + 6}{6} \geqslant 1$

4 Mary went to the record shop with £20. She bought two CDs costing £x each and a DVD costing £9.50. When she got to the till, she found she didn't have enough money.

Mary took the DVD back and paid for the two CDs.

She counted her change and found she had enough money to buy a lipstick for £7.

   a Show that $2x + 9.5 > 20$ and solve the inequality.

   b Show that $2x + 7 \leqslant 20$ and solve the inequality.

   c Show the solution to both of these inequalities on a number line.

   d The price of a CD is a whole number of pounds. How much is a CD?

5 On copies of the number lines below, draw two inequalities so that only the integers {5, 6, 7, 8} are common to both inequalities.

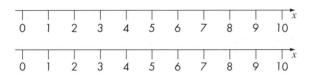

6 Identify x from these three pieces of information.

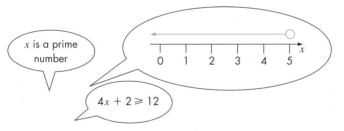

x is a prime number

$4x + 2 \geqslant 12$

7 Solve the following inequalities and illustrate their solutions on number lines.

   a $\dfrac{5x + 2}{2} > 3$ **b** $\dfrac{3x - 4}{5} \leqslant 1$ **c** $\dfrac{4x + 3}{2} \geqslant 11$ **d** $\dfrac{2x - 5}{4} < 2$

   e $\dfrac{8x + 2}{3} \leqslant 2$ **f** $\dfrac{7x + 9}{5} > -1$ **g** $\dfrac{x - 2}{3} \geqslant -3$ **h** $\dfrac{5x - 2}{4} \leqslant -1$

15.7 Graphical inequalities

Homework 15J

1   **a** Draw the line $y = 3$ (as a solid line).   **b** Shade the region defined by $y \geqslant 3$.

2   **a** Draw the line $x = -1$ (as a dashed line).   **b** Shade the region defined by $x < -1$.

3   **a** Draw the line $x = -1$ (as a dashed line).

    b Draw the line $x = 3$ (as a solid line) on the same grid.

    c Shade the region defined by $-1 < x \leqslant 3$.

4 **a** On the same grid, draw the regions defined by these inequalities.

   i $-2 \leqslant x \leqslant 2$ **ii** $-1 < y \leqslant 3$

  b Are the following points in the region defined by both inequalities?

   i $(2, 2)$ **ii** $(-2, 2)$ **iii** $(-2, -1)$ **iv** $(-2, 3)$

5 **a** Draw the line $y = 2x + 1$ (as a solid line).

  b Shade the region defined by $y \leqslant 2x + 1$.

6 **a** Draw the line $3x + 4y = 12$ (as a dashed line).

  b Shade the region defined by $3x + 4y > 12$.

7 **a** Draw the line $y = 2x - 1$ (as a solid line).

  b Draw the line $x + y = 5$ (as a solid line) on the same diagram.

  c Shade the diagram to show the region defined by both $y \geqslant 2x - 1$ and $x + y \leqslant 5$.

8 On the same grid, draw the regions defined by the following inequalities.

  a $x \leqslant 2$ **b** $y \geqslant x - 2$ **c** $x + y \geqslant -2$

9 **a** On the same grid, draw the regions defined by the following inequalities.

   i $y < x + 2$ **ii** $y \geqslant 2x - 2$ **iii** $y \geqslant 0$

  b Are the following points in the region defined by all three inequalities?

  i $(0, 2)$ **ii** $(0, -2)$ **iii** $(2, 2)$ **iv** $(4, 4)$

10 **a** On the same grid, draw the regions defined by the following inequalities.

   i $x \leqslant 2$ **ii** $y > 1$ **iii** $y \leqslant x + 1$

  b Write down the coordinates of all the points whose coordinates are integers and lie in the region that satisfies all the inequalities in part **a**.

11 The graph shows three points, $(1, 1)$, $(2, 1)$ and $(2, 2)$.

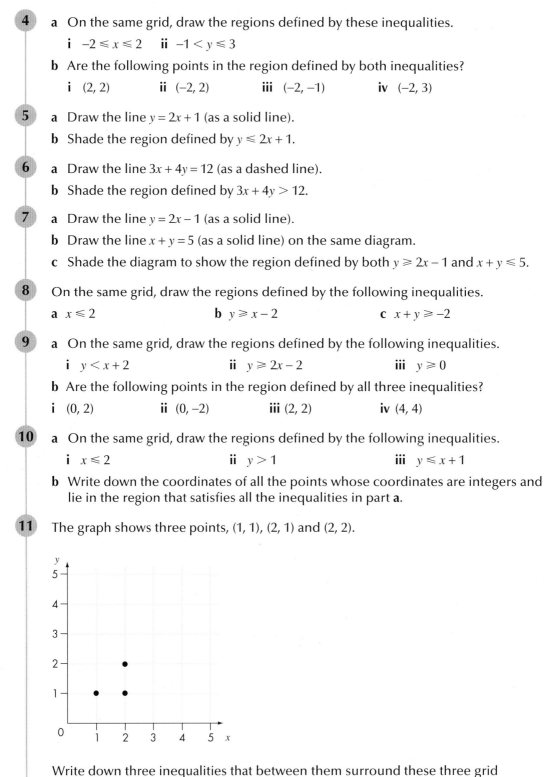

Write down three inequalities that between them surround these three grid intersection points and no others.

15.8 Trial and improvement

Homework 15K 🔢

1. Find two consecutive whole numbers between which the solution to each of the following equations lies.

   a $x^3 + x = 7$ **b** $x^3 + x = 55$ **c** $x^3 + x = 102$ **d** $x^3 + x = 89$

2. Use trial and improvement to find a solution to each equation correct to 1 decimal place.

   a $x^3 - x = 30$ **b** $x^3 - x = 95$ **c** $x^3 - x = 150$ **d** $x^3 - x = 333$

3. Show that $x^3 + x = 45$ has a solution between $x = 3$ and $x = 4$ and use trial and improvement to find the solution correct to 1 decimal place.

4. Show that $x^3 - 2x = 95$ has a solution between $x = 4$ and $x = 5$ and use trial and improvement to find the solution correct to 1 decimal place.

5. Jake is using trial and improvement to solve the equation $x^3 - x^2 = 25$.

   The table shows his first trial.

x	$x^3 - x^2$	Comment
3	18	Too low

   Continue the table to find a solution to the equation, correct to 1 decimal place,.

6. A rectangle has an area of 200 cm². Its length is 8 cm longer than its width. Find, correct to 1 decimal place, the dimensions of the rectangle.

7. Ellen wants to create a rectangular lawn. She wants it to be 15 m longer than it is wide and she wants the area of the lawn to be 800 m².

   Use trial and improvement to find, correct to 1 decimal place, the dimensions for the lawn?

8. A triangle has a vertical height 2 cm longer than its base length. Its area is 20 cm². Find, correct to 1 decimal place, the dimensions of the triangle.

9. The height of a rectangular picture is 3 cm less than its length. Its area is 120 cm². Find, correct to 1 decimal place, the dimensions of the picture.

10. This cuboid has a volume of 1000 cm³.

    a Write down an expression for the volume of the cuboid.

    b Use trial and improvement to find the value of x correct to 1 decimal place.

    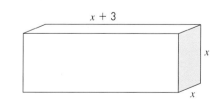

11. Two numbers a and b are such that $ab = 20$ and $a - b = 5$.

    Use trial and improvement to find the two numbers correct to 1 decimal place.

16 Number: Counting, accuracy, powers and surds

16.1 Rational numbers, reciprocals, terminating and recurring decimals

Homework 16A

1 Work out each fraction as a terminating decimal or recurring decimal, as appropriate.

    a $\frac{3}{4}$ **b** $\frac{1}{15}$ **c** $\frac{1}{25}$ **d** $\frac{1}{11}$ **e** $\frac{1}{20}$

2 There are several patterns to be found in recurring decimals. For example,

$$\frac{1}{13} = 0.076\ 923\ 076\ 923\ 076\ 923\ 076\ 923\ldots$$

$$\frac{2}{13} = 0.153\ 846\ 153\ 846\ 153\ 846\ 153\ 846\ldots$$

$$\frac{3}{13} = 0.230\ 769\ 230\ 769\ 230\ 769\ 230\ 769\ldots$$

    a Write down the decimals for $\frac{4}{13}, \frac{5}{13}, \frac{6}{13}, \frac{7}{13}, \frac{8}{13}, \frac{9}{13}, \frac{10}{13}, \frac{11}{13}, \frac{12}{13}$ to 24 decimal places.

    b What do you notice?

3 Write each fraction as a decimal. Use your results to write the list in order of size, smallest first.

    $\frac{2}{9}$ $\frac{1}{5}$ $\frac{23}{100}$ $\frac{2}{7}$ $\frac{3}{11}$

4 Convert each terminating decimal to a fraction.

    a 0.57   **b** 0.275   **c** 0.85   **d** 0.06   **e** 3.65

5 Show that the reciprocal of 1 is 1.

6   **a** Work out the reciprocal of the reciprocal of 4.

    b Work out the reciprocal of the reciprocal of 5.

    c What do you notice?

7   **a** Give an example to show that the reciprocal of a number greater than 1 is less than 1.

    b Give an example to show that the reciprocal of a number less than 0 is also less than 0.

8 $x = 0.024\ 242\ 424\ldots$

    a What is $100x$?

    b By subtracting the original value from your answer to part **a**, work out the value of $99x$.

    c Multiply both sides by 10 to get $990x$ and eliminate the decimal on the right-hand side.

    d Divide both sides by 990.

    e What is x as a fraction in its simplest form?

9 Convert each recurring decimal to a fraction.

    a $0.\dot{7}$ **b** $0.5\dot{7}$ **c** $0.\dot{5}\dot{4}$ **d** $0.2\dot{7}\dot{5}$

    e $2.\dot{5}$ **f** $2.\dot{3}\dot{6}$ **g** $0.0\dot{6}\dot{3}$ **h** $2.0\dot{7}\dot{5}$

10 **a** Write $1.\dot{7}$ as a rational number in the form $\frac{a}{b}$, where a and b are whole numbers.

    b Given that $n = 1.\dot{7}$:

       i write down the value of $10n$

      ii write down the value of $9n$

     iii express n as a rational number, in the form $\frac{a}{b}$, where a and b are whole numbers.

16.2 Estimating powers and roots

Homework 16B

1 Write down the answers to these.

    a $\sqrt{256}$ **b** $\sqrt[3]{1\,000}$ **c** $\sqrt{8^2 + 15^2}$ **d** $\sqrt{15^2 - 6^3}$

2 A square number and two cube numbers have a sum of 156. What are the numbers?

3 Between which two integers does the square root of 150 lie?

4 Between which two integers does the cube root of 380 lie?

5 Which of the following statements are true?

    a All square numbers end in 1, 4, 9 or 6.

    b The product of two numbers is always larger than either of the two numbers.

    c Cube numbers can end in any digit except 9.

    d The product of two numbers is always larger than both the numbers.

    e The square of a number is always greater than the number.

    f Adding two numbers and then squaring them gives the same result as squaring the numbers and then adding them.

    g All whole numbers are either cube numbers or the sum of 2, 3, 4 or 5 cube numbers.

6 Pythagorean triples are sets of three integers a, b and c such that $a^2 + b^2 = c^2$.

    These formulae generate Pythagorean triples when n and m are integers and $n > m$.

       $a = n^2 - m^2$ $b = 2nm$ $c = n^2 + m^2$

    a Choose numbers for n and m and show that the formula works.

    b Show algebraically that $(n^2 - m^2)^2 + (2nm)^2 = (n^2 + m^2)^2$

7 **a** Estimate the value of each number. Do not use a calculator.

       i $\sqrt{6900}$ **ii** $\sqrt{55}$ **iii** $\sqrt[3]{60}$ **iv** 2.8^4

    b Use a calculator to check your answers.

16.3 Negative and fractional powers

Homework 16C

1 You are given that $6^4 = 1296$. Write down the value of 6^{-4}.

2 Write down each number in fraction form.

    a 5^{-2} **b** 4^{-1} **c** 10^{-3} **d** 3^{-3} **e** x^{-2} **f** $5t^{-1}$

3 Write down each number in negative index form.

    a $\dfrac{1}{2^4}$ **b** $\dfrac{1}{7}$ **c** $\dfrac{1}{x^2}$

4 Change each expression into an index form of the type shown.

    a Of the form 2^n **i** 32   **ii** $\dfrac{1}{4}$

    b Of the form 10^n **i** 10 000   **ii** $\dfrac{1}{100}$

    c Of the form 5^n **i** 625   **ii** $\dfrac{1}{125}$

5 Find the value of each of number for the value of the letter shown.

    a $x = 3$ **i** x^2 **ii** $4x^{-1}$

    b $t = 5$ **i** t^{-2} **ii** $5t^{-4}$

    c $m = 2$ **i** m^{-3} **ii** $4m^{-2}$

6 Calculate the value of these expressions when $a = 3$ and $b = 2$. Give each answer as a fraction in its simplest form.

    a $3a^{-1} + 2b^{-2}$ **b** $a^{-2} + b^{-3}$

7 a and b are integers.

    $2^a + 3^b = 41$

    Work out the values of a and b.

8 c and d are integers. c is even.

    Is $c^2 + d^3$ always even, always odd or could be either odd or even.

    Give examples to show how you decided.

9 Put the numbers:

    x^0 x^{-1} x^1

    in order from smallest to largest, when:

    a when x is greater than 1   **b** when x is between 0 and 1

    c when x is between -1 and 0

Homework 16D

1 Evaluate each number.

    a $36^{\frac{1}{2}}$ **b** $144^{\frac{1}{2}}$ **c** $25^{\frac{1}{2}}$ **d** $196^{\frac{1}{2}}$

    e $8^{\frac{1}{3}}$ **f** $125^{\frac{1}{3}}$ **g** $32^{\frac{1}{5}}$ **h** $144^{\frac{1}{2}}$ **i** $27^{\frac{1}{3}}$

2   **a** $\left(\dfrac{25}{81}\right)^{\frac{1}{2}}$ **b** $\left(\dfrac{81}{36}\right)^{\frac{1}{2}}$ **c** $\left(\dfrac{36}{64}\right)^{\frac{1}{2}}$ **d** $\left(\dfrac{8}{27}\right)^{\frac{1}{3}}$

    e $\left(\dfrac{16}{625}\right)^{\frac{1}{4}}$ **f** $\left(\dfrac{4}{9}\right)^{-\frac{1}{2}}$ **g** $\left(\dfrac{16}{25}\right)^{-\frac{1}{2}}$ **h** $\left(\dfrac{8}{27}\right)^{-\frac{1}{3}}$

3 Which of these is the odd one out?

$27^{-\frac{1}{3}}$ $25^{-\frac{1}{2}}$ 3^{-1}

Show how you decided.

4 Imagine that you are the teacher. Write down how you would explain to the class that $16^{-\frac{1}{4}}$ is equal to $\frac{1}{2}$.

5 Find values for x and y that make this equation work.

$x^{-\frac{1}{4}} = y^{-\frac{1}{2}}$

Homework 16E

1 Evaluate each expression.

a $16^{\frac{3}{4}}$ **b** $125^{\frac{4}{3}}$ **c** $81^{\frac{3}{4}}$

2 Rewrite each number in index form.

a $\sqrt[4]{t^3}$ **b** $\sqrt[5]{m^2}$

3 Evaluate each expression.

a $27^{\frac{2}{3}}$ **b** $8^{\frac{4}{3}}$ **c** $36^{\frac{3}{2}}$ **d** $81^{1.25}$

4 Use trial and improvement method, or another method that you prefer, to solve these equations.

a $6^x = 60$ **b** $10^x = 2$

5 **a** Evaluate $8^{\frac{1}{3}}$.   **b** Write $16^{-\frac{1}{2}} \times 2^{-3}$ as a power of 2.

c Given that $32^y = 2$, find the value of y.

6 Which of these is the odd one out?

$16^{-\frac{3}{4}}$ $64^{-\frac{1}{2}}$ $8^{-\frac{2}{3}}$

Show how you decided.

7 Imagine that you are the teacher. Write down how you would explain to the class that $27^{-\frac{2}{3}}$ is equal to $\frac{1}{9}$.

16.4 Surds

Homework 16F

1 Simplify each expression. Leave your answers in surd form if necessary.

a $\sqrt{3} \times \sqrt{4}$ **b** $\sqrt{5} \times \sqrt{7}$ **c** $\sqrt{5} \times \sqrt{5}$ **d** $\sqrt{2} \times \sqrt{32}$

2 Simplify each expression. Leave your answers in surd form if necessary.

a $\sqrt{15} \div \sqrt{5}$ **b** $\sqrt{18} \div \sqrt{2}$ **c** $\sqrt{32} \div \sqrt{2}$ **d** $\sqrt{12} \div \sqrt{8}$

3 Simplify each expression. Leave your answers in surd form if necessary.

a $\sqrt{3} \times \sqrt{3} \times \sqrt{2}$ **b** $\sqrt{5} \times \sqrt{5} \times \sqrt{15}$ **c** $\sqrt{2} \times \sqrt{8} \times \sqrt{8}$ **d** $\sqrt{2} \times \sqrt{8} \times \sqrt{5}$

4 Simplify each expression. Leave your answer in surd form if necessary.

a $\sqrt{3} \times \sqrt{8} \div \sqrt{2}$ **b** $\sqrt{15} \times \sqrt{3} \div \sqrt{5}$ **c** $\sqrt{8} \times \sqrt{8} \div \sqrt{2}$ **d** $\sqrt{3} \times \sqrt{27} \div \sqrt{3}$

5 Simplify each surd into the form $a\sqrt{b}$.

   a $\sqrt{90}$ **b** $\sqrt{32}$ **c** $\sqrt{63}$ **d** $\sqrt{300}$

   e $\sqrt{150}$ **f** $\sqrt{270}$ **g** $\sqrt{96}$ **h** $\sqrt{125}$

6 Simplify each expression.

   a $2\sqrt{32} \times 5\sqrt{2}$ **b** $4\sqrt{8} \times 2\sqrt{2}$ **c** $4\sqrt{12} \times 5\sqrt{3}$ **d** $3\sqrt{6} \times 2\sqrt{6}$

   e $2\sqrt{5} \times 5\sqrt{3}$ **f** $2\sqrt{3} \times 3\sqrt{3}$ **g** $2\sqrt{2} \times 3\sqrt{8}$ **h** $2\sqrt{3} \times 2\sqrt{27}$

   i $8\sqrt{24} \div 2\sqrt{3}$ **j** $3\sqrt{27} \div \sqrt{3}$ **k** $5\sqrt{18} \div \sqrt{2}$ **l** $2\sqrt{32} \div 4\sqrt{8}$

   m $5\sqrt{2} \times \sqrt{8} \div 2\sqrt{2}$ **n** $3\sqrt{15} \times \sqrt{3} \div \sqrt{5}$ **o** $2\sqrt{24} \times 5\sqrt{3} \div 2\sqrt{8}$

7 Find the value of a that makes each equation true.

   a $\sqrt{5} \times \sqrt{a} = 20$ **b** $\sqrt{3} \times \sqrt{a} = 12$ **c** $\sqrt{5} \times 4\sqrt{a} = 20$

8 Simplify theses expression.

   a $\left(\dfrac{\sqrt{2}}{3}\right)^2$ **b** $\left(\dfrac{4}{\sqrt{3}}\right)^2$

9 Simplify these expressions.

   a $\sqrt{32} + \sqrt{8}$ **b** $\sqrt{32} \times \sqrt{8}$ **c** $\sqrt{27} \times \sqrt{18} \div \sqrt{3}$

10 Decide whether this statement is true or false.

   $\sqrt{(a^2 + b^2)} = a + b$

   Show your working.

11 Write down a division expression containing two different surds that has an integer answer.

Homework 16G

1 Rationalise the denominators of each expression.

   a $\dfrac{1}{\sqrt{7}}$ **b** $\dfrac{1}{\sqrt{8}}$ **c** $\dfrac{2}{\sqrt{5}}$ **d** $\dfrac{1}{2\sqrt{2}}$

   e $\dfrac{5\sqrt{3}}{\sqrt{27}}$ **f** $\dfrac{\sqrt{8}}{\sqrt{3}}$ **g** $\dfrac{1 + \sqrt{3}}{\sqrt{3}}$ **h** $\dfrac{3 - \sqrt{2}}{\sqrt{8}}$

2 Show that each statement is true:

   a $(3 + \sqrt{5})(2 + \sqrt{5}) = 11 + 5\sqrt{5}$ **b** $(3 - \sqrt{2})(3 + \sqrt{2}) = 7$

3 Expand and simplify where possible.

   a $\sqrt{5}(3 - \sqrt{2})$ **b** $\sqrt{8}(3 - 4\sqrt{2})$ **c** $3\sqrt{8}(2\sqrt{2} + 4)$

   d $(2 + \sqrt{3})(1 - \sqrt{3})$ **e** $(3 + \sqrt{5})(2 - \sqrt{5})$ **f** $(3 - \sqrt{2})(4 + 2\sqrt{2})$

4 Work out the value of x in these triangles. Give your answers in the simplest possible form.

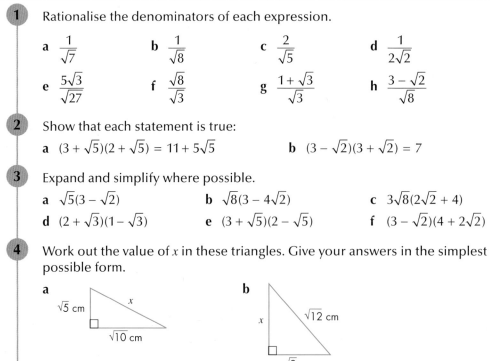

   a $\sqrt{5}$ cm, x, $\sqrt{10}$ cm

   b $\sqrt{12}$ cm, x, $\sqrt{8}$ cm

5 Calculate the area of these rectangles. Simplify your answers where possible.

a
1 + √2 cm
2 − √2 cm

b
2 + √7 cm
√3 cm

6 Expand and simplify each expression

a $(5 + \sqrt{3})(5 - \sqrt{3})$ **b** $(6 - \sqrt{2})(6 + \sqrt{2})$

7 **a** Write down two surds that, when added together, give a rational number.

b Write down two surds that, when added together, do not give a rational number.

8 Katie is working out the height of a curtain for a window in metres. Her calculator displays the answer $1 + \sqrt{3}$.

Without using a calculator, show that the height of the curtain is between 2 metres and 3 metres.

16.5 Limits of accuracy

Homework 16H

1 Write down error interval of each measurement.

a 5 cm measured to the nearest centimetre

b 50 mph measured to the nearest 10 mph

c 15.2 kg measured to the nearest tenth of a kilogram

d 75 km measured to the nearest 5 km

2 Round these numbers to the given degree of accuracy.

a 45.678 to 3 sf   **b** 19.96 to 2 sf   **c** 0.3213 to 2 dp

3 Write down the limits of accuracy for each measurement. Each is rounded to the given degree of accuracy.

a 7 cm (1 sf)   **b** 18 kg (2 sf)   **c** 30 min (2 sf)   **d** 747 km (3 sf)

e 9.8 m (1 dp)   **f** 32.1 kg (1 dp)   **g** 3.0 h (1 dp)   **h** 90 g (2 sf)

i 4.20 mm (2 dp)   **j** 2.00 kg (2 dp)   **k** 34.57 min (2 dp)   **l** 100 m (2 sf)

4 Write down the lower and upper bounds of each measurement. Each is given to the accuracy stated.

a 6 m (1 sf)   **b** 34 kg (2 sf)   **c** 56 min (2 sf)   **d** 80 g (2 sf)

e 3.70 m (2 dp)   **f** 0.9 kg (1 dp)   **g** 0.08 s (2 dp)   **h** 900 g (2 sf)

i 0.70 m (2 dp)   **j** 360 d (3 sf)   **k** 17 weeks (2 sf)   **l** 200 g (2 sf)

5 A theatre has 365 seats.

For a show, 280 tickets are sold in advance.

The theatre manager estimates that another 100 people, to the nearest 10, will turn up without tickets and that 5% of those with tickets will not turn up.

If she is correct, is it possible they will all get a seat?

Show clearly how you decide.

6 A parking space is 4.8 m long, measured to the nearest tenth of a metre.

A car is 4.5 metres long, measured to the nearest half a metre.

Which of the following statements is definitely true?

A: The space is big enough.

B: The space is not big enough.

C: It is impossible to tell whether or not the space is big enough.

Explain how you decide.

7 1 litre = 100 cl

A carton contains 1 litre of milk, to the nearest 10 cl.

What is the least amount of milk in the carton?

Give your answer in centilitres.

8 Natasha has 20 identical bricks. Each brick is 15 cm long, measured to the nearest centimetre.

a What is the greatest possible length of one brick?

b What is the smallest possible length of one brick?

c If the bricks are put end to end, what is the greatest possible length of all the bricks?

d If the bricks are put end to end, what is the least possible length of all the bricks?

9 A whole number, when rounded to 2 significant figures, is 450. When rounded to 1 significant figure it is 400. What is the range of values for the number?

16.7 Problems involving limits of accuracy

Homework 16I

1 Cans have a mass of 250 g, to the nearest 10 g.

What are the minimum and maximum masses of ten of these cans?

2 The cans in question **1** are stacked on a shelf that can safely hold 15 kg.

What is the maximum number of cans that can definitely be stacked safely on the shelf?

3 A crate of the cans in question **1** has a mass of 24 kg.

What are the minimum and maximum numbers of cans that could be in the crate?

4 These are the dimensions of rectangles. In each case, find the limits of accuracy of the area. The measurements are shown to the level of accuracy indicated in brackets.

a 3 cm × 8 cm (nearest cm)   **b** 3.2 cm × 6.4 cm (1 dp)

c 7.86 cm × 18.78 cm (2 dp)

5 A rectangular garden has sides of 8 m and 5 m, measured to the nearest metre.

a Write down the limits of accuracy for each length.

b What is the maximum area of the garden?

c What is the minimum perimeter of the garden?

6 A playground is measured as 32 m by 45 m, to the nearest metre. Calculate the limits of accuracy for the area of the playground.

7 The measurements of a box, to the nearest centimeter, are given as 12 cm by 8 cm by 5 cm. Calculate the limits of accuracy for the volume of the box.

8 Miss Hoffman is a plumber.

She has a 10-metre length of pipe, correct to the nearest metre.

She uses 2 m of pipe on her first job, 3 m on the second job and 4 m on his third job. Each measurement is to the nearest half metre.

What is the longest length of pipe she could have left?

9 Belinda is going to walk for 5 miles.

She walks at an average speed of 3 mph, to the nearest whole number of miles per hour.

If she sets off at 2 pm, what is the latest time that she will finish the walk?

10 The area of a rectangular field is given as 400 m², to the nearest 10 m². One length is given as 24 m, to the nearest metre. Find the limits of accuracy for the other length of the field.

11 In triangle ABC, AB = 8 cm, BC = 6 cm and ∠ABC = 42°. All the measurements are given to the nearest unit. Calculate the limits of accuracy for the area of the triangle.

12 A stopwatch records the time for the winner of a 100-metre race as 12.3 seconds, measured to the nearest one-tenth of a second.

a What are the greatest and least possible times for the winner?

b The length of the 100-metre track is correct to the nearest centimetre. What are the greatest and least possible lengths of the track?

c What is the fastest possible average speed of the winner?

13 A cube has a volume of 27 cm³, to the nearest cubic centimetre. Find the range of possible values of the side length of the cube.

14 A cube has a volume of 125 cm³, to the nearest cubic centimetre. Find the limits of accuracy of the area of one side of the square base.

15 In the triangle ABC, the length of side AB is 42 cm, to the nearest centimetre. The length of side AC is 35 cm, to the nearest centimeter and angle C is 61°, to the nearest degree. What is the largest possible value of angle B?

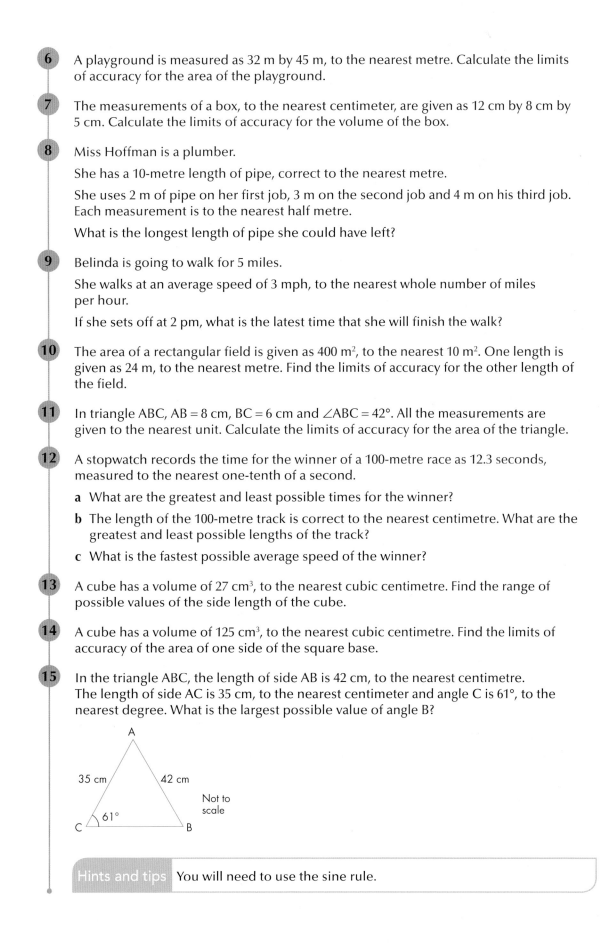

Not to scale

Hints and tips You will need to use the sine rule.

16.8 Choices and outcomes

Homework 16J

1 How many numbers between 0 and 200 have at least one digit of 7?

2 Use your calculator to work out these numbers.

    a 8!   **b** 30!

3 The pin codes for a cash point are four digits selected from a choice of ten (0–9). How many possible codes are there?

4 Use your calculator to work these out.   a $_6P_3$ b $_8C_4$

5 Use your calculator to work these out.   a $_5P_3$ b $_7C_4$

6 A combination lock has four wheels. Each wheel has the digits 0 to 9 on it.

    a How many different combinations are possible?

    b Fred has forgotten his combination. He knows it uses the four digits of his house number, which is 6152. How many possible combinations will Fred need to try to be certain he finds the correct one?

7 16 runners take part in a 10 000-metre race. How many different ways can the first three places be filled?

8   **a** Two cards are taken from a regular 52-card pack, *with* replacement. What is the probability that a Jack is drawn, followed by a Queen?

    b Two cards are taken from a regular 52-card pack, *without* replacement. What is the probability that a Jack is drawn, followed by a Queen?

9 How many four-letter arrangements are there of the letters in the word POLO?

10 A box contains four red balls, two blue balls and three yellow balls. If at least one blue ball must be included, how many different colour arrangements can be made of three balls from the box?

17 Algebra: Quadratic equations

17.1 Plotting quadratic graphs

Homework 17A

1 **a** Copy and complete the table or use a calculator to work out values for the graph of $y = 2x^2$ for values of x from -3 to 3.

x	-3	-2	-1	0	1	2	3
$y = 2x^2$	18		2			8	

b Use your graph to find the value of y when $x = -1.4$.

c Use your graph to find the values of x that give a y-value of 10.

2 **a** Copy and complete the table or use a calculator to work out values for the graph of $y = x^2 + 3$ for $-5 \leq x \leq 5$.

x	-5	-4	-3	-2	-1	0	1	2	3	4	5
$y = x^2 + 3$	28		12					7			28

b Use your graph to find the value of y when $x = 2.5$.

c Use your graph to find the values of x that give a y-value of 10.

3 **a** Copy and complete the table or use a calculator to work out values for the graph of $y = x^2 - 3x + 2$ for $-3 \leq x \leq 4$.

x	-3	-2	-1	0	1	2	3	4
$y = x^2 - 3x + 2$	20			2			2	

b Use your graph to find the y- value when $x = -1.5$.

c Use your graph to find the values of x that give a y-value of 2.5.

4 Tom is drawing quadratic equations of the form $y = x^2 + bx + c$.

He notices that two of his graphs pass through the point (2, 5).

He has drawn the graphs of two of the following equations. Which two?

Equation A:	$y = x^2 + 3$
Equation B:	$y = x^2 + 1$
Equation C:	$y = x^2 + 2x - 3$
Equation D:	$y = x^2 - x + 5$

17.2 Solving quadratic equations by factorisation

Homework 17B

1 Solve these equations.

a $(x + 3)(x + 2) = 0$ **b** $(t + 4)(t + 1) = 0$ **c** $(a + 5)(a + 3) = 0$

d $(x + 4)(x - 1) = 0$ **e** $(x + 2)(x - 5) = 0$ **f** $(t + 3)(t - 4) = 0$

g $(x - 2)(x + 1) = 0$ **h** $(x - 1)(x + 4) = 0$ **i** $(a - 6)(a + 5) = 0$

j $(x - 2)(x - 5) = 0$ **k** $(x - 2)(x - 1) = 0$ **l** $(a - 2)(a - 6) = 0$

2 First factorise, then solve each of these equations.

a $x^2 + 6x + 5 = 0$ **b** $x^2 + 9x + 18 = 0$ **c** $x^2 - 7x - 8 = 0$

d $x^2 - 4x - 21 = 0$ **e** $x^2 + 3x - 10 = 0$ **f** $x^2 + 2x - 15 = 0$

g $t^2 - 4t - 12 = 0$ **h** $t^2 - 3t - 18 = 0$ **i** $x^2 + x - 2 = 0$

j $x^2 - 4x + 4 = 0$ **k** $m^2 - 10m + 25 = 0$ **l** $t^2 - 10t + 16 = 0$

m $t^2 + 7t + 12 = 0$ **n** $k^2 - 3k - 18 = 0$ **o** $a^2 - 20a + 64 = 0$

3 Ella is x years old.

Her brother is four years older than her.

The product of their ages is 1020.

a Set up a quadratic equation to represent this situation.

b How old is Ella?

4 A rectangular field is 140 m longer than it is wide.

A combine harvester cuts corn at the rate of 200 m² per minute.

It takes four hours to cut the field.

What are the dimensions of the field?

Homework 17C

Give your answers either in rational form or as mixed numbers.

1 Here are three equations.

A: $(x - 2)^2 = 0$ **B:** $3x - 2 = 4$ **C:** $x^2 - 4x + 4 = 0$

a Give a mathematical fact that equations A and C have in common.

b Give a mathematical fact that equations A, B and C have in common.

2 Pythagoras' theorem states that the sum of the squares of the two short sides of a right-angled triangle equals the square of the long side (hypotenuse).

A right-angled triangle has sides $2x$, $2x + 1$ and $x + 1$ cm.

a Show that $x^2 - 2x = 0$

b Find the area of the triangle.

17 Algebra: Quadratic equations

3　**a** Solve these equations.

　　i $2x^2 + 5x + 2 = 0$　　　　**ii** $7x^2 + 8x + 1 = 0$　　　　**iii** $4x^2 + 3x - 7 = 0$

　　iv $6x^2 + 13x + 5 = 0$　　　**v** $6x^2 + 7x + 2 = 0$

　b Rearrange these equations into the general form and then solve them.

　　i $x^2 - x = 6$　　　　　　**ii** $2x(4x + 7) = -3$　　　　**iii** $(x + 3)(x - 4) = 18$

　　iv $11x = 21 - 2x^2$　　　　**v** $(2x + 3)(2x - 3) = 9x$

　c **i** Simplify $\dfrac{x^2 - 9}{3x - 9}$.　　　**ii** Solve the equation $12x^2 - 25x + 12 = 0$.

17.3 Solving a quadratic equation by using the quadratic formula

Homework 17D 🖩

1　Use the quadratic formula to solve these equations. Give your answers to 2 decimal places.

　a $3x^2 + x - 5 = 0$　　　　**b** $2x^2 + 4x + 1 = 0$　　　　**c** $x^2 - x - 7 = 0$

　d $3x^2 + x - 1 = 0$　　　　**e** $3x^2 + 7x + 3 = 0$　　　　**f** $2x^2 + 11x + 1 = 0$

　g $2x^2 + 5x + 1 = 0$　　　　**h** $x^2 + 2x - 9 = 0$　　　　**i** $x^2 + 2x - 6 = 0$

2　A rectangular lawn is 5 m longer than it is wide.

　The area of the lawn is 60 m².

　How long is the lawn? Give your answer to the nearest centimetre.

3　Gerard is solving a quadratic equation using the formula.

　He correctly substitutes values for a, b and c to get:

　　$x = \dfrac{4 \pm \sqrt{112}}{6}$

　What is the equation that Gerard is trying to solve?

4　Eric uses the quadratic formula to solve $9x^2 - 12x + 4 = 0$.

　Anna uses factorisation to solve the same equation.

　They both find something unusual in their solutions.

　Explain what this is, and why.

5　Solve the equation $x^2 = 5x + 7$. Give your answers correct to 3 significant figures.

Homework 17E

1　Work out the discriminant $b^2 - 4ac$ of the equations. In each case say how many solutions the equation has.

　a $x^2 + x + 3 = 0$　　　　**b** $3x^2 - 4x + 2 = 0$　　　　**c** $x^2 - 6x - 12 = 0$

　d $8x^2 + 8x + 2 = 0$　　　　**e** $36x^2 - 9x = 0$　　　　**f** $4x^2 - 9 = 0$

2 A quadratic equation has the answers $x = 3 \pm 6\sqrt{7}$. The coefficient of x^2 is 1. Find the value of $b^2 - 4ac$.

3 Poppy works out the discriminant of the quadratic equation $x^2 + bx - c = 0$ as $b^2 - 4ac = 17$.

There are four possible equations that could lead to this discriminant. What are they?

4 For which values of k does the equation $x^2 + (k + 1)x + (4 - k) = 0$ have only one answer? Give the values in surd form.

17.4 Solving quadratic equations by completing the square

Homework 17F

1 Write an equivalent expression in the form $(x \pm a)^2 - b$.

a $x^2 + 10x$ **b** $x^2 + 18x$ **c** $x^2 - 8x$ **d** $x^2 + 20x$ **e** $x^2 + 7x$

2 Write an equivalent expression in the form $(x \pm a)^2 - b$.

a $x^2 + 10x - 1$ **b** $x^2 + 18x - 5$ **c** $x^2 - 8x + 3$ **d** $x^2 - 5x - 1$

3 Solve these equations by completing the square. Leave your answers in surd form where appropriate.

a $x^2 + 10x - 1 = 0$ **b** $x^2 + 18x - 5 = 0$ **c** $x^2 - 8x + 3 = 0$

d $x^2 + 20x + 7 = 0$ **e** $x^2 - 5x - 1 = 0$

4 Solve $x^2 + 8x - 3 = 0$ by completing the square. Give your answers to 2 decimal places.

5 **a** Write the equation $x^2 + 4x - 6$ in the form $(x + a)^2 - b$.

b Hence or otherwise, solve the equation $x^2 + 4x - 6 = 0$. Give your answer in surd form.

6 Dhiaan rewrites the expression $x^2 + px + q$ by completing the square.

She correctly does this and gets $(x + 3)^2 - 17$.

What are the values of p and q?

7 The following statements are steps in the method of completing the square to solve the equation $x^2 + 12x - 11 = 0$. Rearrange the steps to give a logical method.

A $x = -6 \pm \sqrt{47}$

B $(x + 6)^2 - 47 = 0$

C $(x + 6)^2 - 36 - 11 = 0$

D $(x + 6)^2 = 47$

E $x + 6 = \pm \sqrt{47}$

17.5 The significant points of a quadratic curve

Homework 17G

1 **a** Copy and complete the table.

x	−1	0	1	2	3	4	5	6
$y = x^2 - 5x + 4$	10	4				0		

b Draw the graph of $y = x^2 - 5x + 4$ for $-1 \le x \le 6$ and use it to find the roots of the equation $x^2 - 5x + 4 = 0$.

2 **a** Copy and complete the table.

x	−1	0	1	2	3	4	5
$y = x^2 - 3x + 2$	6	2			1		

b Draw the graph of $y = x^2 - 3x + 2$ for $-1 \le x \le 5$ and use it to find the roots of the equation $x^2 - 3x + 2 = 0$.

3 **a** Copy and complete the table.

x	−5	−4	−3	−2	−1	0	1	2
$y = x^2 + 4x - 6$	−1							6

b Draw the graph of $y = x^2 + 4x - 6$ for $-5 \le x \le 2$ and use it to find the roots of the equation $x^2 + 4x - 6 = 0$.

4 Look at your answers to question 1. Write down:

a the coordinates of the point where the graph crosses the y-axis

b the coordinates of the turning point of the graph.

5 Look at your answers to question 2. Write down:

a the coordinates of the point where the graph crosses the y-axis

b the coordinates of the turning point of the graph.

6 Look at your answers to question 3.

a Write down the coordinates of the turning point of the graph.

b Write the equation $x^2 + 4x - 6 = 0$ in the form $(x - a)^2 + b = 0$.

c What is the connection between minimum point and the values in the equation when it is written as $(x - a)^2 + b$.

d Without drawing the curve, predict the turning point of the graph $y = x^2 + 6x - 5$.

Homework 17H

1 **a** Plot the graph of $y = x^2 - 4x - 21$ for $-5 \le x \le 10$.

b Write down the coordinates of:

   i the y-intercept   **ii** the points where the curve intersects the x-axis

   iii the turning point.

2 Work out the roots and the y-intercept of each graph.

   a $y = x^2 - 16$ **b** $y = x^2 - 8x$ **c** $y = x^2 - 5x - 6$

3 Work out the coordinates of the turning point of the graph of $y = x^2 - 8x + 2$.

4 Work out the minimum value of the expression $x^2 - 6x + 5$.

5 **a** Work out the turning point of the graph of $y = x^2 + 2x + 1$.

   b What does your answer tell you about the roots of $y = x^2 + 2x + 1$?

6 Sketch the graph of $y = 2x^2 - 17x + 8$. You should include the roots, y-intercept and turning point.

7 Melvin draws a quadratic graph that has a minimum point at $(5, -10)$ but forgets to label it.

   He knows it is of the form $y = x^2 + px + q$.

   Help Melvin to find values of p and q.

8 Beryl draws a quadratic graph which has a minimum point at $(2, -6)$ but forgets to label it.

   She knows that it is of the form $y = x^2 + px + q$.

   Help Beryl to find the values of p and q.

17.6 Solving one linear and one non-linear equation using graphs

Homework 17I

1 Using graphical methods to find the approximate or exact solutions to the following pairs of simultaneous equations. In this question, suitable ranges for the axes are given in brackets.

   a $y = x^2 + 5x - 3$ and $y = x$ $(-10 \leq x \leq 5, -10 \leq y \leq 5)$

   b $x^2 + y^2 = 25$ and $x + y = 2$ $(-6 \leq x \leq 6, -6 \leq y \leq 6)$

   c $y = x^2 - 3x + 2$ and $y = x + 2$ $(-5 \leq x \leq 5, -5 \leq y \leq 5)$

   d $y = x^2 - 5$ and $y = x + 3$ $(-5 \leq x \leq 5, -6 \leq y \leq 10)$

2 **a** Solve the simultaneous equations $y = x^2 + 2x - 1$ and $y = 4x - 2$ $(-5 \leq x \leq 5, -5 \leq y \leq 10)$.

   b What is special about the intersection of these two graphs?

   c Show that $4x - 2 = x^2 + 2x - 1$ can be rearranged to $x^2 - 2x + 1 = 0$.

   d Factorise and solve $x^2 - 2x + 1 = 0$.

   e Explain how the solution in part **d** relates to the intersection of the graphs.

3 **a** Solve the simultaneous equations $y = x^2 + 3x + 5$ and $y = 2x - 1$ $(-5 \leq x \leq 5, -5 \leq y \leq 8)$.

   b What is special about the intersection of these two graphs?

   c Rearrange $2x - 1 = x^2 + 3x + 5$ into the general quadratic form $ax^2 + bx + c = 0$.

   d Work out the discriminant $b^2 - 4ac$ for the quadratic in part **c**.

   e Explain how the value of the discriminant relates to the intersection of the graphs.

17.7 Solving quadratic equations by the method of intersection

Homework 17J

1 The graph is of $y = x^2 - 2x - 4$.

Solve these equations.

a $x^2 - 2x - 4 = 0$

b $x^2 - 2x - 4 = 4$

c $x^2 - 2x - 3 = 0$

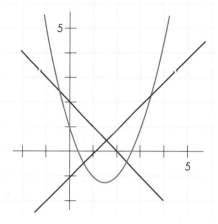

2 The graphs of $y = x^2 - 3x + 1$, $y = x - 1$ and $y + x = 2$ are drawn on the axes shown.

Solve these equations.

a $x^2 - 3x + 1 = 0$

b $2x - 3 = 0$

c $x^2 - 3x - 1 = 0$

d $x^2 - 4x + 2 = 0$

e $x^2 - 2x - 1 = 0$

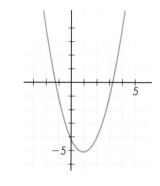

3 Draw the graph of $y = x^3 - 2x + 3$.

a Use the graph to solve these equations.

   i $x^3 - 2x + 3 = 0$ **ii** $x^3 - 2x = 0$

b By drawing a suitable straight line, solve $x^3 - 3x + 2 = 0$.

4 The graph is of $y = x^3 - 4x - 1$.

a Solve these equations.

   i $x^3 - 4x - 1 = 0$ **ii** $x^3 - 4x + 2 = 0$

b By drawing a suitable straight line, solve $x^3 - 5x - 1 = 0$.

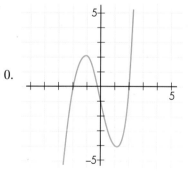

5 The graph is of $y = x^3 - 4x$.

  a Find two positive solutions for $x^3 - 4x = -2$.

  b By drawing a suitable straight line, solve $x^3 - 3x + 1 = 0$.

6 The graph shows the lines
A: $y = x^2 + 5x - 3$; B: $y = x$;
C: $y = x + 3$; D: $y + x = 2$ and
E: $y = -x$.

  a Which pair of lines has a common solution of $(-1.5, 1.5)$?

  b Which pair of lines has the approximate solutions $(1, 1)$ and $(-6.8, 8.8)$?

  c What quadratic equation has approximate solutions of $(-5.2, -2.2)$ and $(1.2, 4.2)$?

  d The minimum point of the graph $y = x^2 + 5x - 3$ is at $(-2.5, -9.25)$.

  Write down the minimum point of the graph $y = x^2 + 5x - 8$.

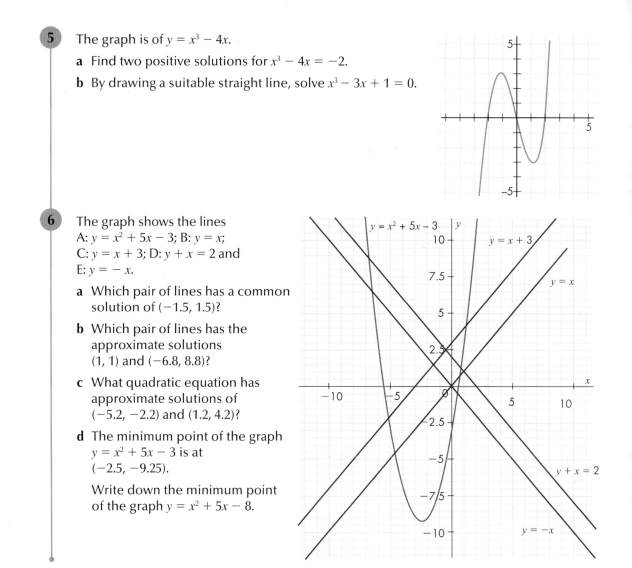

17.8 Solving linear and non-linear simultaneous equations algebraically

Homework 17K

1 Solve each pair of simultaneous equations.

  a $xy = 3$
  $y = x - 2$

  b $xy = 2$
  $2y - x = 3$

  c $x^2 + y^2 = 29$
  $x - y = 7$

  d $y = x^2 + 3x + 4$
  $y = 1 - x$

  e $y = x^2 - 3x + 5$
  $y = 2x - 1$

  f $y = 4x^2 + 2x + 1$
  $y = 3x^2 + 2x + 2$

2 **a** Solve the simultaneous equations $y = x^2 + 2x - 5$ and $y = 6x - 9$.

b Which of the sketches below represents the graph of the equations in **a**?
Explain your choice.

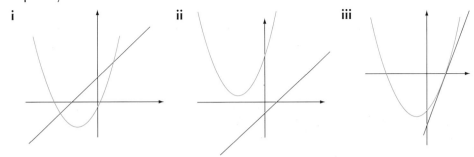

i   **ii**   **iii**

3 The solutions of the simultaneous equations $y = x^2 - 3x - 3$ and $y = ax + b$ are
(6, 15) and (−1, 1).

Find the values of a and b.

17.9 Quadratic inequalities

Homework 17L

1 Solve these inequalities.

a $x^2 - 25 > 0$ **b** $x^2 - 81 \leq 0$ **c** $x^2 - x < 0$ **d** $4x^2 - 16x \geq 0$

2 State all the integers which satisfy these inequalities.

a $x^2 - 25 \leq 0$ **b** $x^2 - 8x + 15 \leq 0$

3 Solve these inequalities.

a $x^2 - 4x - 12 > 0$ **b** $x^2 + 14x + 45 < 0$

c $5x^2 - 13x - 6 \geq 0$ **d** $3x^2 \leq 5x + 12$

4 Solve these inequalities and illustrate their solutions on number lines.

a $x^2 + 5x - 14 < 0$ **b** $x^2 + 9x + 20 > 0$

5 Find the set of values of x that satisfy both $5x - 10 \leq 3(x - 4)$ and $x^2 + 5x + 6 < 0$.

6 A rectangle has sides of $(x - 3)$ m and $(x - 2)$ m.

Its perimeter is greater than 14 m and its area is less than 56 m².

Represent the possible values of x on a number line.

18 Statistics: Sampling and more complex diagrams

18.1 Collecting data

Homework 18A

1 Describe how you could use the data-handling cycle to test each hypothesis
  Remember to include the type of data (primary or secondary) that you would use in
  each case.

  a January is the coldest month of the year.

  b Girls are better than boys at estimating weights.

  c More men go to cricket matches than women.

  d The TV show *Strictly Come Dancing* is watched by more women than men.

  e The older you are the more likely you are to go ballroom dancing.

2 You have been asked to make a presentation of the timing of the school day and so
  decide to interview a sample of students. Show how you will choose students to
  interview to ensure your results are representative and reliable. Give reasons for any
  decisions you make.

3 Comment on the reliability of these data samples.

  a Finding out about religious beliefs in your school by asking the year 11 Religious
    Education option class.

  b Finding out how many homes have microwaves by asking the first 100 students
    who walk through the school gates.

  c Finding out the most popular Playstation game by asking a Year 7 form to choose
    their favourite game.

4 Comment on the suitability of these samples. For any that are not satisfactory, suggest
  a more reliable sampling method.

  a Bill wanted to find out what proportion of his school went to the cinema, so he
    obtained an alphabetical list of students and sent a questionnaire to every tenth
    student on the list.

  b The council wants to know about sports facilities in an area so they sent a survey
    team to the local shopping centre one Monday morning.

  c A political party wanted to know how much support they had in an area so they
    rang 500 people from the phone book in the evening.

5 A2B Trains wanted to estimate how many people in a certain town used their train
  services. They telephoned 200 people in the town one evening and asked: "Have you
  travelled by train in the last week?" Thirty-two people said "Yes". From this, A2B Trains
  concluded that 16% of the town's population used their services.

  Give three criticisms of this method of estimation.

6 Gwen wants to find out the opinions of students in her school. She chooses to interview 150 students.

a Explain why this is a suitable size of sample.

b The table shows the number of students in each year group in the school.

Complete a similar table showing the types of students that should be in Gwen's sample.

Year	Boys	Girls	Total
7	154	137	291
8	162	156	318
9	134	160	294
10	153	156	309
11	130	140	270
Total	**733**	**749**	**1482**

7 You are asked to conduct a survey at a concert where the attendance is approximately 15 000.

Explain how you could take a sample of the crowd.

8 Mrs Reynolds, the deputy head at Bradway School, wanted to find out how often the sixth form students left the school building to eat lunch. The table shows the number of students in the two year groups of the sixth form.

	Boys	Girls
Y12	88	92
Y13	83	75

a Design a short questionnaire that Mrs Reynolds could use to sample the school.

b Mrs Reynolds wanted to use a sample of 50 students. How many of each group of students should complete the questionnaire?

18.2 Frequency polygons

Homework 18B

1 The table shows the number of goals scored by a football team in 20 matches.

Goals	0	1	2	3	4
Frequency	5	7	4	3	1

a Draw a frequency polygon to illustrate the data.

b Calculate the mean number of goals scored per game.

2 The table shows the times taken by 50 students to complete a multiplication square.

Time, s (seconds)	$10 < s \leqslant 20$	$20 < s \leqslant 30$	$30 < s \leqslant 40$	$40 < s \leqslant 50$	$50 < s \leqslant 60$
Frequency	4	10	16	12	8

a Draw a frequency polygon to illustrate the data.

b Calculate an estimate for the mean time taken by the students.

3 A supermarket manager wants to ensure that the average queuing time for customers is no more than 5 minutes. The table shows the results of a survey of customer queuing times.

Time, m (minutes)	$0 < m \leqslant 2$	$2 < m \leqslant 4$	$4 < m \leqslant 6$	$6 < m \leqslant 8$	$8 < m \leqslant 10$
Frequency	3	5	10	8	4

a Draw a frequency polygon to illustrate the data.

b Calculate an estimate for the mean customer queuing time.

c What advice would you give the manager about average waiting times?

4 The frequency polygon shows the lengths of time that students spent watching TV during a week.

Calculate an estimate of the mean time spent watching TV by the students.

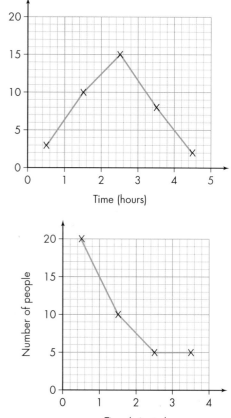

5 The frequency polygon shows the times that a number of people waited at a set of temporary traffic lights one morning on their way to work.

Dave said, "Most people spent 30 seconds waiting."

Show that he may be wrong.

6 The table shows the results of a mental arithmetic test, collated separately for girls and boys.

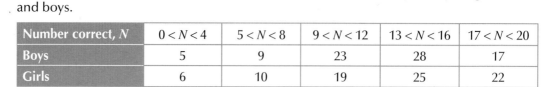

Number correct, N	$0 < N < 4$	$5 < N < 8$	$9 < N < 12$	$13 < N < 16$	$17 < N < 20$
Boys	5	9	23	28	17
Girls	6	10	19	25	22

a On the same axes, draw frequency polygons to illustrate the boys' scores and the girls' scores.

b Estimate the mean score for boys and girls scores separately.

c Comment on your results.

18.3 Cumulative frequency graphs

Homework 18C

1 An army squad all undertook a timed one-mile run. The table shows the results recorded by the trainer.

    a Copy the table and complete the cumulative frequency column.

    b Draw a cumulative frequency graph.

    c Use your graph to estimate the median time and the interquartile range.

Time, t (seconds)	Frequency	Cumulative frequency
$200 < t \leqslant 240$	3	
$240 < t \leqslant 260$	7	
$260 < t \leqslant 280$	12	
$280 < t \leqslant 300$	23	
$300 < t \leqslant 320$	7	
$320 < t \leqslant 340$	5	
$340 < t \leqslant 360$	5	

2 A company had 360 web pages. They monitored and recorded the number of visits during 24 hours.

    a Copy the table and complete a cumulative frequency column.

    b Draw a cumulative frequency graph.

    c Use your graph to estimate the median number of visits to one of the company's web pages and the interquartile range.

    d The company decides to rewrite pages with fewer than 60. Approximately how many pages will they rewrite?

Number of visits, v	Frequency
$0 < v \leqslant 50$	6
$50 < v \leqslant 100$	9
$100 < v \leqslant 150$	15
$150 < v \leqslant 200$	25
$200 < v \leqslant 250$	31
$250 < v \leqslant 300$	37
$300 < v \leqslant 350$	32
$350 < v \leqslant 400$	17
$400 < x \leqslant 450$	5

3 Two classes complete two papers – Paper 1 and Paper 2 – for their class exams. The head of year plotted cumulative frequency graphs for the results.

    a What is the median score for each paper?

    b What is the interquartile range for each paper?

    c Which is the harder paper? Show how you know.

    d The teachers wanted 90% of the students to pass each paper and 15% of the students to get top marks in each paper.

    What marks for each paper give:

    i a pass   **ii** the top grade?

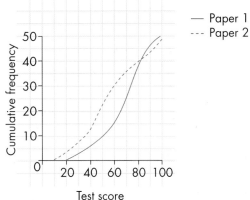

4 Johnny was given a cumulative frequency graph showing the results of a number of students in a spelling test. He was told the top 15% were given the top grade.

Explain how you would find the marks needed to gain this top grade?

18.4 Box plots

Homework 18D

1 The box plot shows the number of peas per pod for a crop grown by a prize-winning gardener.

A young gardener also grew a crop of peas. Her peas per pod were: least number 3, lower quartile 4.75, median 5.5, upper quartile 6.25, highest number 9.

a Copy the diagram. On the same grid, draw a box plot for the young gardener's crop.

b Comment on the differences between the two distributions.

2 The box plot shows the monthly salaries of all the men in a computer firm.

The data for the women in the company is lowest salary £600, lower quartile £1300, median 1600, upper quartile £2000, highest salary £2400.

a Copy the diagram and on the same grid draw a box plot for the women's salaries.

b Comment on the differences between the two distributions.

3 The box plots show the hours of life of two brands of batteries.

a Comment on the differences in the two distributions.

b Ricardo wants to buy some batteries for his calculator. Which brand would you recommend and why?

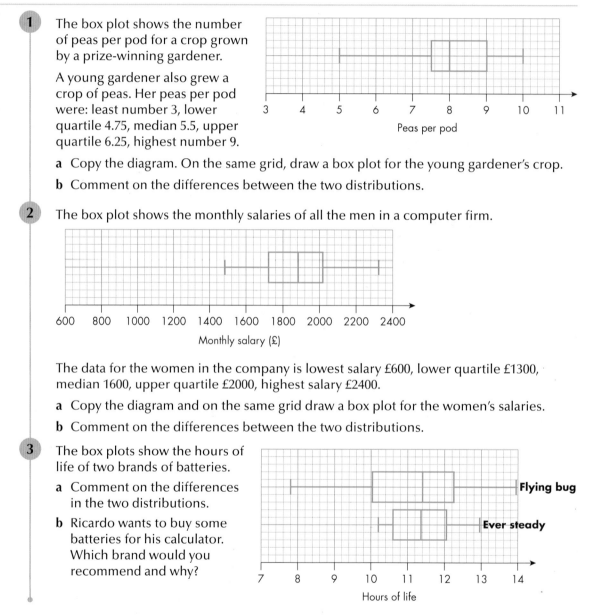

4 The table shows some data about the durations of telephone calls to two operators working on a mobile phone helpline.

	Shortest duration	Lower quartile	Median duration	Upper quartile	Longest duration
Justin	1 m 10 s	2 m 20 s	3 m 30 s	4 m 50 s	7 m 10 s
Julia	40 s	2 m 20 s	5 m 10 s	7 m 30 s	10 m 45 s

a Draw box plots to compare the sets of data.

b Comment on the differences between the distributions.

c The company has reduce its workforce and can only keep one of these operators. Who should they let go and why?

5 A school entered 80 students for an examination. The results are shown in the table.

Mark, x	$0 < x \leqslant 20$	$20 < x \leqslant 40$	$40 < x \leqslant 60$	$60 < x \leqslant 80$	$80 < x \leqslant 100$
Number of students	2	14	28	26	10

a Calculate an estimate for the mean.

b Complete a cumulative frequency table for the data and draw a cumulative frequency graph.

c i Use your graph to estimate the median mark.

　ii 12 of these pupils were given a grade A. Use your graph to estimate the lowest mark for which grade A was given.

d Another school also entered 80 students for the same examination. Their results were lowest mark 40, lower quartile 50, median 60, upper quartile 70, highest mark 80. Draw a box plot to compare the sets of data. Comment on the differences between the two distributions.

6 A dental practice had two dentists, Dr Ball and Dr Charlton.

The practice manager drew these box plots to illustrate the waiting times for their patients during November.

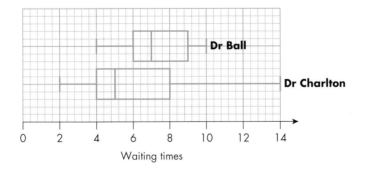

Gabrielle was deciding which dentist to try to see. Which one would you advise she chooses? Give reasons to support your answer.

7 Joy was given a diagram showing box plots for the daily hours of sunshine in the seaside resorts of Scarborough and Blackpool for July. No scale was shown.

She was told to write a report on the differences between the amounts of sunshine in both resorts.

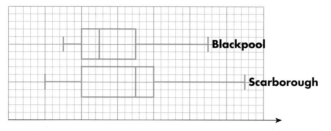

Write a report that she could produce from these box plots without a scale.

8 These are the box plots for a school's end-of-year science examination.

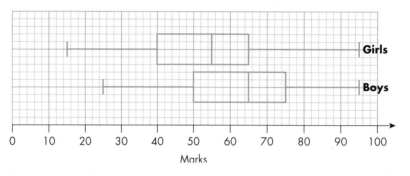

Estimate the difference between the means of the boys' and the girls' examination marks.

18.5 Histograms

Homework 18E

1 The table shows the waiting times for customers at a supermarket checkout.

a Draw a histogram of for this data.

b Calculate an estimate for the mean waiting time.

Waiting time (minutes)	Frequency
$0 < x \leqslant 2$	15
$2 < x \leqslant 4$	7
$4 < x \leqslant 6$	12
$6 < x \leqslant 8$	15
$8 < x \leqslant 10$	12

2 **a** The grouped frequency table shows the ages of 300 people watching a film at the cinema. Draw a histogram to show the data.

Age, x years	$0 < x \leqslant 10$	$10 < x \leqslant 20$	$20 < x \leqslant 30$	$30 < x \leqslant 40$	$40 < x \leqslant 50$
Frequency	25	85	115	45	30

b This grouped frequency table shows the distribution of ages at a different film.

Age, x years	$20 < x \leqslant 30$	$30 < x \leqslant 40$	$40 < x \leqslant 50$	$50 < x \leqslant 60$	$60 < x \leqslant 70$
Frequency	35	120	130	50	15

Draw a histogram to show this data.

c Comment on the differences between the distributions.

3 **a** The table shows the ages of 300 people at a cinema.

Age, x years	$0 < x \leqslant 20$	$20 < x \leqslant 30$	$30 < x \leqslant 50$
Frequency	110	115	75

Draw a histogram to show the data.

b Compare your histogram to the one you drew in question **1a**. What do you notice?

4 The histogram shows the journey time to school of a group of students.

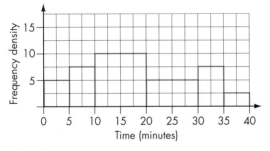

a Copy and complete the table.

Time (minutes)	$0 < x \leqslant 5$	$5 < x \leqslant 10$	$10 < x \leqslant 20$	$20 < x \leqslant 30$	$30 < x \leqslant 35$	$35 < x \leqslant 40$
Frequency						

b Calculate an estimate for the mean of the distribution.

5 The table shows the waiting times for customers at a supermarket checkout.

Waiting time (minutes)	Frequency
$0 < x \leqslant 1$	15
$1 < x \leqslant 3$	7
$3 < x \leqslant 4$	12
$4 < x \leqslant 5$	15
$5 < x \leqslant 10$	12

a Draw a histogram for this data.

b Calculate an estimate for the median waiting time. Show your working.

6 Andrew was asked to create a histogram.

Explain to Andrew how he can find the height of each bar on the frequency density scale.

7 The histogram shows the scores from a school's end of year science tests.

a Estimate the median score.

b Estimate the interquartile range of the scores.

c Calculate an estimate for the mean score.

d Given that 10% of the students gained an A, what was the lowest score needed for an A?

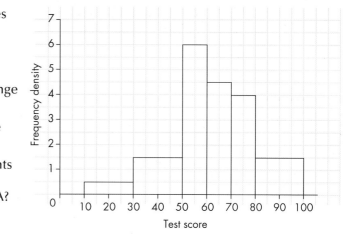

8 The histogram shows the distances travelled to church by members of its congregation.

22 members travel between **4** km and **6** km to church. What is the probability of choosing a member at random who travels more than **4** km to church?

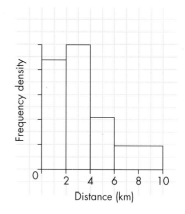

19 Probability: Combined events

19.1 Addition rules for outcomes of events

Homework 19A

1 Dami throws an ordinary dice. What is the probability that he throws:

   a an even number   **b** a 5   **c** an even number or a 5?

2 Jane picks a card from a standard pack. What is the probability that she picks:

   a a red card   **b** a black card   **c** a red or a black card?

3 Natalie picks a card from a standard pack. What is the probability that she picks:

   a an ace   **b** a king   **c** an ace or a king?

4 Januz chooses a card at random from this set.

S T A T I S T I C S

What is:

   a P(choosing an S)   **b** P(choosing a vowel)   **c** P(choosing an S or a vowel)?

5 A bag contains ten white balls, twelve black balls and eight red balls. A ball is picked from the bag at random. What is the probability that it will be:

   a white   **b** black   **c** black or white

   d not red   **e** not red or black?

6 A spinner is numbered and coloured as shown in the diagram. The probabilities are given in the tables.

Colour	Probability
Orange	0.5
Yellow	0.25
Pink	0.25

Number	Probability
1	0.4
2	0.35
3	0.25

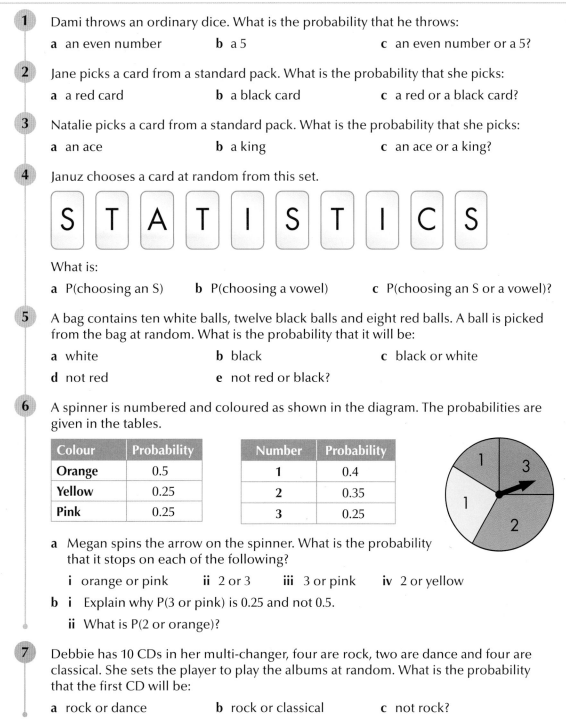

   a Megan spins the arrow on the spinner. What is the probability that it stops on each of the following?

      i orange or pink   **ii** 2 or 3   **iii** 3 or pink   **iv** 2 or yellow

   b i Explain why P(3 or pink) is 0.25 and not 0.5.

      ii What is P(2 or orange)?

7 Debbie has 10 CDs in her multi-changer, four are rock, two are dance and four are classical. She sets the player to play the albums at random. What is the probability that the first CD will be:

   a rock or dance   **b** rock or classical   **c** not rock?

8 Frank buys one dozen free-range eggs. The farmer tells him that a quarter of the eggs his hens lay have double yolks.

    a How many eggs with double yolks can Frank expect to get?

    b He cooks three and finds they all have a single yolk. He argues that he now has a 1 in 3 chance of a double yolk from the remaining eggs. Give reasons why he is wrong.

9 John has a bag containing six red, five blue and four green balls. He picks one ball from the bag at random. What is the probability that the ball is:

    a red or blue   **b** not blue   **c** pink   **d** red or not blue?

10 Poppy and Ben create a playlist on their MP3 player so they can have a variety of background music at a dinner party. The list has 100 different tracks: 10 love songs, 15 rap tracks, 35 rock tracks and 40 contemporary tracks.

    They will set it to play the tracks continuously, at random.

    a What is the probability that:

        i the first track played is a love song

        ii the last track is either rock or a contemporary track

        iii the track playing when they start their meal is not a rap track?

    b They plan to announce their engagement at midnight and want to have either a love song or a contemporary track playing. What is the probability that they will not get a track of their choice?

    c The party will last for six hours. For what amount of time, in hours and minutes, would you expect the MP3 player to be playing rock tracks?

11 Joy, Vicky and Max play cards together every Sunday night. Joy is always the favourite to win, with a probability of 0.65.

    In 2014 there were 52 Sundays and Vicky won 10 times.

    How many times in the year would you expect Max to have won?

12 Kathy has blue, black, brown and yellow jumpers in her wardrobe.

    She asks her brother to throw a jumper from her wardrobe down the stairs to her.

    Show that P(neither blue nor yellow) might not be the same as P(not blue) + P(not yellow)?

19.2 Combined events

Homework 19B

1 Kirsty throws two fair dice, each numbered from 1 to 6. Her score is the sum of the numbers on the dice.

    a Draw a sample space diagram to show the total score.

    b What is the probability that her score is:

        i 7   **ii** 5 or 8   **iii** bigger than 9

        iv from 2 to 5 inclusive   **v** odd   **vi** not a square number?

2 Draw a probability diagram to show the pairs of outcomes when two fair dice are thrown, for example, (3, 3).

What is the probability that:

a the score is a 'double'

b at least one of the dice shows a 3

c the score on one dice is three times the score on the other dice

d at least one of the dice shows an odd number

e both dice show a 5

f at least one of the dice will show a 5

g exactly one dice shows a 5?

3 Two fair dice are thrown. The score on the first dice is doubled and the score on the second dice is subtracted from it.

a Copy and complete the sample space diagram.

b Write down:

   i P(1)   **ii** P(a negative number)

   iii P(an even number)   **iv** P(0 or 1)

   v P(a prime number).

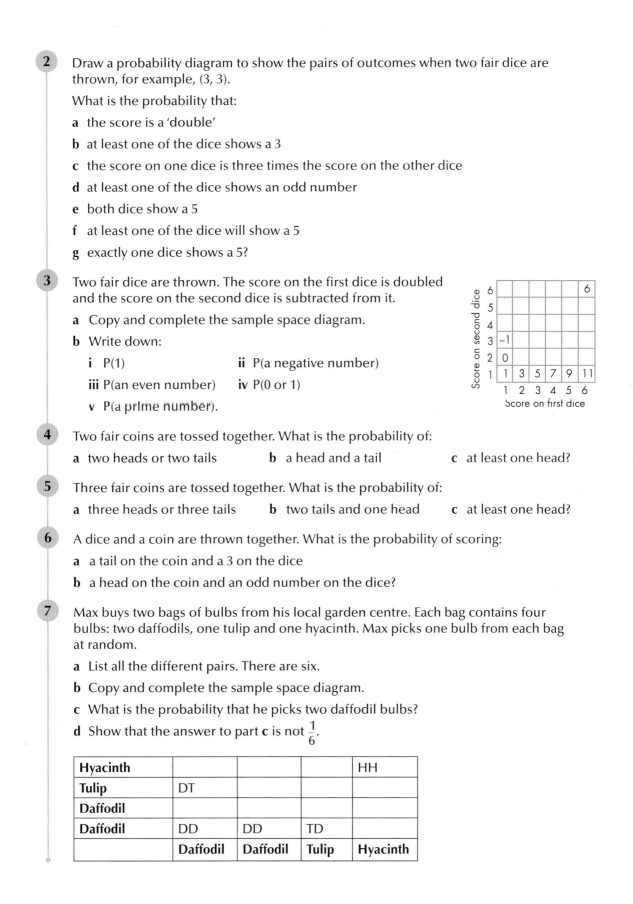

4 Two fair coins are tossed together. What is the probability of:

a two heads or two tails   **b** a head and a tail   **c** at least one head?

5 Three fair coins are tossed together. What is the probability of:

a three heads or three tails   **b** two tails and one head   **c** at least one head?

6 A dice and a coin are thrown together. What is the probability of scoring:

a a tail on the coin and a 3 on the dice

b a head on the coin and an odd number on the dice?

7 Max buys two bags of bulbs from his local garden centre. Each bag contains four bulbs: two daffodils, one tulip and one hyacinth. Max picks one bulb from each bag at random.

a List all the different pairs. There are six.

b Copy and complete the sample space diagram.

c What is the probability that he picks two daffodil bulbs?

d Show that the answer to part **c** is not $\frac{1}{6}$.

Hyacinth				HH
Tulip	DT			
Daffodil				
Daffodil	DD	DD	TD	
	Daffodil	Daffodil	Tulip	Hyacinth

8 Shehab walked into his local supermarket and saw a sign for a competition.

Roll 2 dice!
Score a total of 2 and win a £20 note.
Only 50p a go.

Shehab thought about having a go.

a Draw the sample space diagram for this event.

b What is the probability of winning a £20 note?

c How many goes should he have in order to expect to win at least one prize?

d If he spent £50 entering the competition, how many times could he expect to win?

9 I throw four coins. What is the probability that I will get more heads than tails?

10 I throw a dice four times and add the four numbers obtained.

Explain the difficulty in drawing a sample space to show all the possible outcomes.

19.3 Tree diagrams

Homework 19C

1 Emmie throws a fair dice twice.

a Copy and complete the tree diagram to show all the possible outcomes.

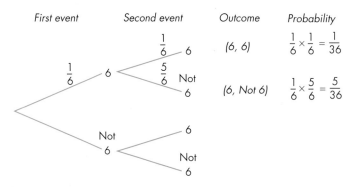

First event	Second event	Outcome	Probability
$\frac{1}{6}$ 6	$\frac{1}{6}$ 6	(6, 6)	$\frac{1}{6} \times \frac{1}{6} = \frac{1}{36}$
	$\frac{5}{6}$ Not 6	(6, Not 6)	$\frac{1}{6} \times \frac{5}{6} = \frac{5}{36}$
Not 6	6		
	Not 6		

b Use the tree diagram to work out the probability of throwing:

i two sixes   **ii** exactly one six   **iii** no sixes.

2 A card is drawn at random from a standard pack of cards. It is replaced, the pack is shuffled and another card is drawn at random.

a What is the probability that either card was a Spade?

b What is the probability that either card was not a Spade?

c Draw a tree diagram to show all the possible outcomes of two cards being drawn as described.

d Use the tree diagram to work out the probability both cards will be Spades.

3 A bag contains three red and two blue balls. A ball is picked at random, replaced, and then another ball is picked at random.

a What is the probability that the first ball picked will be red?

b Copy and complete the tree diagram to show all the possible outcomes.

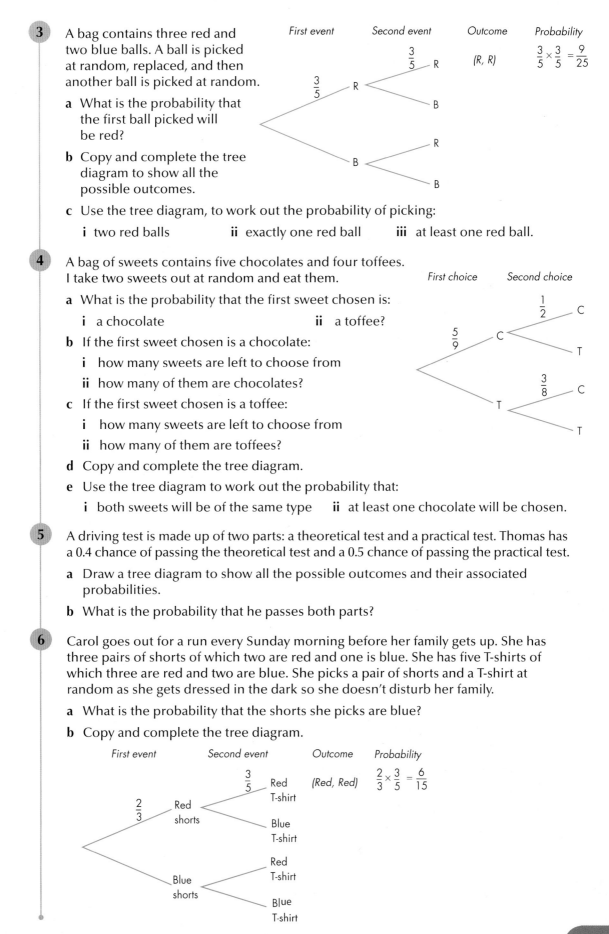

c Use the tree diagram, to work out the probability of picking:

   i two red balls   **ii** exactly one red ball   **iii** at least one red ball.

4 A bag of sweets contains five chocolates and four toffees. I take two sweets out at random and eat them.

a What is the probability that the first sweet chosen is:

   i a chocolate   **ii** a toffee?

b If the first sweet chosen is a chocolate:

   i how many sweets are left to choose from

   ii how many of them are chocolates?

c If the first sweet chosen is a toffee:

   i how many sweets are left to choose from

   ii how many of them are toffees?

d Copy and complete the tree diagram.

e Use the tree diagram to work out the probability that:

   i both sweets will be of the same type   **ii** at least one chocolate will be chosen.

5 A driving test is made up of two parts: a theoretical test and a practical test. Thomas has a 0.4 chance of passing the theoretical test and a 0.5 chance of passing the practical test.

a Draw a tree diagram to show all the possible outcomes and their associated probabilities.

b What is the probability that he passes both parts?

6 Carol goes out for a run every Sunday morning before her family gets up. She has three pairs of shorts of which two are red and one is blue. She has five T-shirts of which three are red and two are blue. She picks a pair of shorts and a T-shirt at random as she gets dressed in the dark so she doesn't disturb her family.

a What is the probability that the shorts she picks are blue?

b Copy and complete the tree diagram.

c What is the probability that Carol goes running in:

    i   shorts and T-shirt that are the same colour

    ii   shorts and a T-shirt that are a different colour

    iii at least one red item?

7 Bob has a bag containing four blue balls, five red balls and one green ball. Sally has a bag containing two blue balls and three red balls. The balls are identical except for the colour. Bob chooses a ball at random from his bag; Sally chooses a ball at random from her bag.

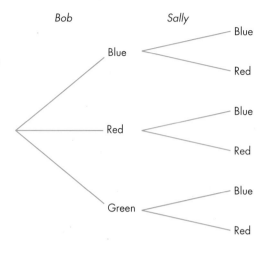

  a Copy the tree diagram and write the probability of each outcome on the appropriate branch.

  b Calculate the probability that both Bob and Sally will choose a blue ball.

  c Calculate the probability that the ball chosen by Bob will be a different colour from the ball chosen by Sally.

8 *Rushdown* is played with a standard pack of cards. In *Rushdown*, you are dealt two cards. If both cards are the numbers 6 or 7 or 8, you have been dealt a 'Tango'.

What is the probability of being dealt a 'Tango'? Give your answer to 3 decimal places.

9 I have a drawer containing black, blue and brown socks. Explain how a tree diagram can help me find the probability of picking, at random, two socks of the same colour.

19.4 Independent events

Homework 19D

1 The probability that a school meal contains turnip on any day is $\frac{1}{4}$. Steve eats in the school canteen five days a week.

  a What is the probability that in five days no meal contains turnips?

  b What is the probability that Steve gets turnip at least once in five days?

2 Lottie throws three fair coins. What is the probability that she throws:

  a three tails   **b** at least one head?

3 Adam is a talented all-round athlete. He is entered for the 100-metre race, the javelin and the high jump. The probability that he wins each of these events is $\frac{4}{5}$, $\frac{3}{4}$ and $\frac{1}{2}$ respectively.

What is the probability that:

  a he doesn't win any of the three events   **b** he wins at least one event?

4 A bag contains seven red and three blue balls. Rajan takes out a ball, notes it colour and replaces it. He then takes out another ball. Work out:

    a P(both balls are red)   **b** P(both balls are blue)   **c** P(at least one ball is red).

5 **a** Cian throws a fair coin three times. What is the probability that he throws:

      i three heads   **ii** no heads   **iii** at least one head?

    b A fair coin is thrown four times. What is the probability of throwing:

      i four heads   **ii** no heads   **iii** at least one head?

    c A fair coin is thrown five times. What is the probability of throwing:

      i five heads   **ii** no heads   **iii** at least one head?

    d A fair coin is thrown n times. What is the probability of throwing:

      i n heads   **ii** no heads   **iii** at least one head?

6 A bag contains two black balls and five red balls. A ball is taken out, its colour noted and replaced. This is repeated three times. What is the probability that:

    a all three are black   **b** exactly two are black

    c exactly one is black   **d** none are black?

7 A bag contains two blue balls and five white balls. A ball is taken out but not replaced. This is repeated three times. What is the probability that:

    a all three are blue   **b** exactly two are blue

    c exactly one is blue   **d** none are blue?

8 A fair dice is thrown three times. What is the probability that:

    a four sixes are thrown   **b** no sixes are thrown   **c** exactly one six is thrown?

9 The probability that Ann is late for school is 0.7. The probability that Bob is late is 0.6. The probability that Cedrick is late is 0.3. Work out:

    a P(exactly one of them is late)   **b** P(exactly two of them are late).

10 Daisy takes three AS exams in Mathematics. The probability that she will pass Pure 1 is 0.9. The probability that she will pass Statistics 1 is 0.65. The probability she will pass Discrete 1 is 0.95. What is the probability that she will pass:

    a all three modules   **b** exactly two modules   **c** at least two modules?

11 Six out of ten cars in Britain use petrol. The rest use diesel. Three cars can be seen approaching in the distance.

    a What is the probability that the first one uses petrol?

    b What is the probability that exactly two of them use diesel?

    c Explain why, if the first car uses petrol, the probability that the second car uses petrol is still 0.6.

12 Six T-shirts are hung out at random on a washing line. Three are red and three are blue. Using R and B for red and blue, write down all 20 possible combinations: for example, RRRBBB, RRBBBR and so on. What is the probability of:

    a three red shirts being next to each other

    b three blue shirts being next to each other?

13 In a class of students, 21 have dark hair, 7 have fair hair and 2 have red hair. Two students are chosen at random to collect in homework.

   a What is the probabilty that they:

     i both have fair hair

     ii do not have the same colour hair?

   b If three students are chosen, what is the probability that exactly two have dark hair?

14 A firm is employing temporary workers. They call 30 for interview and find that seven of them have an HGV licence.

   The firm hires three of the interviewees. What is the probability that:

   a all three have an HGV licence

   b only one has an HGV licence

   c at least two have an HGV licence?

15 Paul was playing a card game and was dealt three cards, all Aces. He thought the chance of him now being dealt another Ace was $\frac{1}{52}$.

   Show why he was wrong.

19.5 Conditional probability

Homework 19E

1 Two unbiased dice with numbered faces are thrown together. The score is the number shown on the uppermost face. What is:

   a P(total score of 8) given that one die shows a 2

   b P(at least one 3) on either dice?

2 Based on previous results, the probability that Manchester United win is $\frac{2}{3}$, the probability they draw is $\frac{1}{4}$ and the probability they lose is $\frac{1}{12}$.

   They play three matches. What is the probability that:

   a they win all three matches   **b** they win exactly two matches

   c they win at least one match?

3 In the Spark'n, Spit'n and Fizz'n machine, the probability that the Spark fails is 0.02. The probability that the Spit fails is 0.08 and the probability that the Fizz fails is 0.05.

   a What is the probability that nothing fails?

   b The machine will still work with one component out of action. What is the probability that it works?

4 On average, Steve is late for school two days each school week.

   a What is the probability that he is late on any one day?

   b In a week of five days, what is the probability that:

     i he is late every day   **ii** he is late exactly once

     iii he is never late   **iv** he is late at least once?

5 The probability that Rob wakes up early at the weekend on either a Saturday or a Sunday but not both is 0.42.

The probability that he wakes up early is the same on both Saturday and Sunday.

What are the two possible answers for the probability of him waking up late on one of the days?

> **Hints and tips** Draw a tree diagram.

6 A bag contains five black balls and three white balls. Three balls are taken out, one by one.

   a If the balls are put back each time, what is the probability of getting:

     i three black balls   **ii** at least one black ball?

   b If the balls are not put back each time, what is the probability of getting:

     i three black balls   **ii** at least one black ball?

7 Two cards are drawn one at a time from a pack of cards and replaced. What is the probability that at least one of them is a heart?

8 Two cards are drawn from a pack of cards without replacement. What is the probability that at least one of them is a heart?

9 A box contains 50 batteries. It is known that 20 of them do not work. John takes out three batteries. What is the probability that:

   a all three of them do not work   **b** at least one of them works?

10 From a box of 100 batteries, Janet takes out three batteries for her radio. The radio will work if two or three of the batteries are good. Given that the probability of a battery from the box working is 0.8, what is the probability that the radio will work?

11 Dan has four blue socks and two black socks in a drawer. He picks out two socks at random. What is the probability that:

   a both socks are blue   **b** both socks are black

   c both socks are the same colour   **d** at least one of the socks is blue?

12 One-ninth of people are left-handed. In a room of five people, what is the probability that:

   a all five are left-handed   **b** all five are right-handed

   c at least one of them is left-handed?

13 In a pack of cards, the aces, kings, queens and jacks are all called 'picture cards'. What is the probability of being dealt four 'picture cards' in a row from a standard pack of cards?

14 A bag of jelly babies contains yellow, green and orange sweets, all the same size. Trishna is asked to find the probability of picking out at random two jelly babies of the same colour.

Describe how Trishna would do this, explaining carefully the point where she is most likely to go wrong.

20.1 Circle theorems

Homework 20A

1 Work out the size of the angle marked x in each circle. O is the centre.

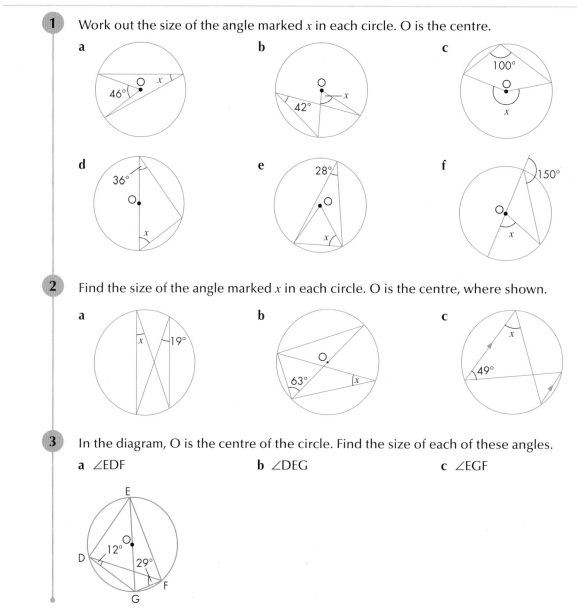

a

46°
O
x

b

O
x
42°

c

100°
O
x

d

36°
O
x

e

28°
O
x

f

150°
O
x

2 Find the size of the angle marked x in each circle. O is the centre, where shown.

a

x 19°

b

O
63°
x

c

x
49°

3 In the diagram, O is the centre of the circle. Find the size of each of these angles.

  a $\angle$EDF   **b** $\angle$DEG   **c** $\angle$EGF

E
O
12°
D
29°
F
G

4 Find the values of x and y in each circle. O is the centre, where shown. Give reasons for your answers.

a

b

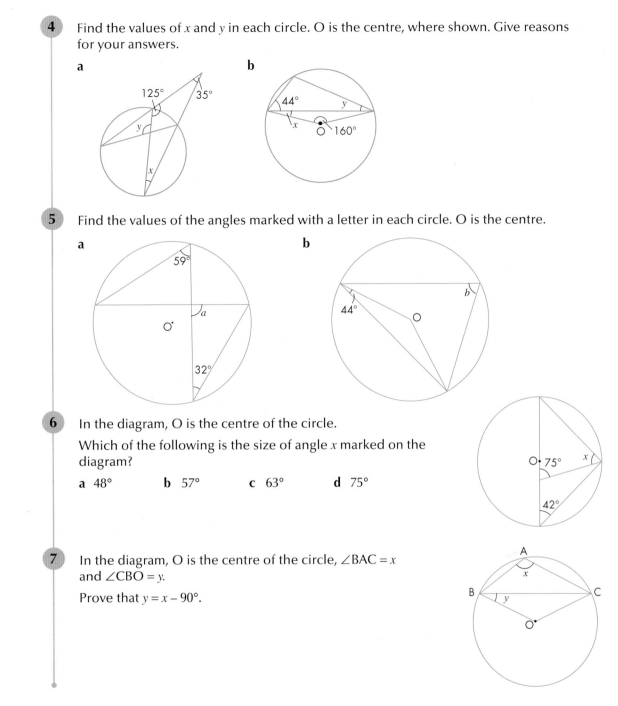

5 Find the values of the angles marked with a letter in each circle. O is the centre.

a

b

6 In the diagram, O is the centre of the circle.

Which of the following is the size of angle x marked on the diagram?

a 48°   **b** 57°   **c** 63°   **d** 75°

7 In the diagram, O is the centre of the circle, $\angle BAC = x$ and $\angle CBO = y$.

Prove that $y = x - 90°$.

20.2 Cyclic quadrilaterals

Homework 20B

1 Find the sizes of the lettered angles in each diagram.

a   **b**   **c**   **d**

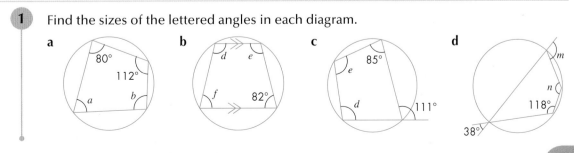

2 Find the sizes of the lettered angles in each diagram.

a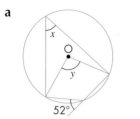

b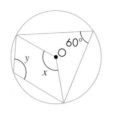

c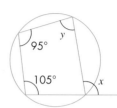

3 Find the sizes of the lettered angles in each diagram. O is the centre of the circle, where shown.

a

b

c

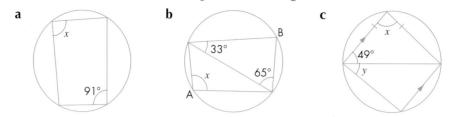

4 ABCD is a cyclic quadrilateral.

Work out the values of x and y.

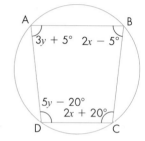

5 In the diagram, O is the centre of the circle.

Show that the angle BOD is 128°. Give reasons for your answer.

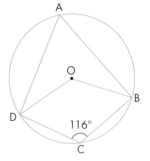

6 A, B, C and D are four points on a circle. AB is parallel to CD.
Prove that $\angle x = \angle y$.

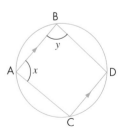

20.3 Tangents and chords

Homework 20C 🔢

1. TP and TQ are tangents to a circle with centre O. Find values for r and x.

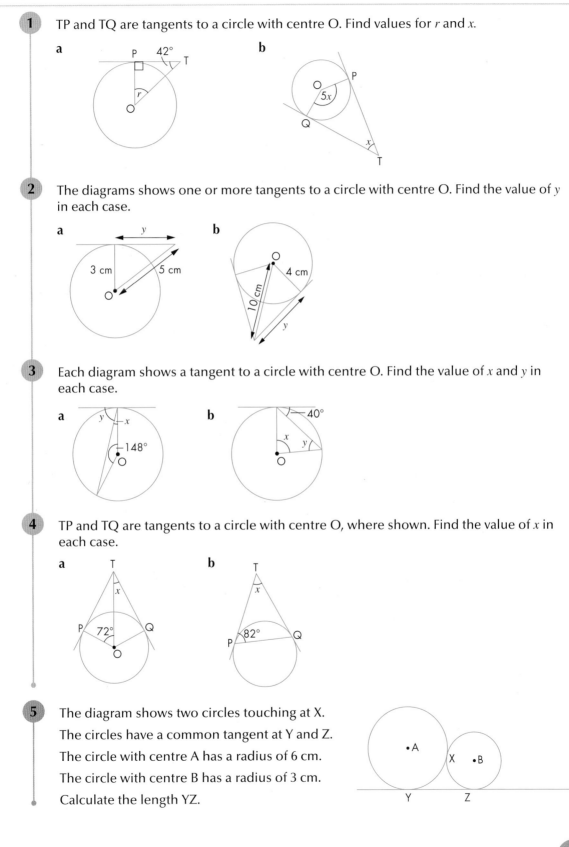

   a

   P   42°   T

   r

   O

   b

   O   P

   $5x$

   Q

   x

   T

2. The diagrams shows one or more tangents to a circle with centre O. Find the value of y in each case.

   a

   y

   3 cm   5 cm

   O

   b

   O

   4 cm

   10 cm

   y

3. Each diagram shows a tangent to a circle with centre O. Find the value of x and y in each case.

   a

   y x

   148°

   O

   b

   40°

   x y

   O

4. TP and TQ are tangents to a circle with centre O, where shown. Find the value of x in each case.

   a

   T

   x

   P   72°   Q

   O

   b

   T

   x

   P   82°   Q

5. The diagram shows two circles touching at X.
   The circles have a common tangent at Y and Z.
   The circle with centre A has a radius of 6 cm.
   The circle with centre B has a radius of 3 cm.
   Calculate the length YZ.

   • A

   X   • B

   Y   Z

6 Two circles intersect at X and Y.

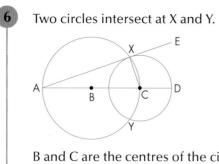

B and C are the centres of the circles and ABCD is a straight line.

Prove that the line AE is a tangent to the smaller circle.

20.4 Alternate segment theorem

Homework 20D

1 Find the sizes of the lettered angles in each diagram.

a   **b**

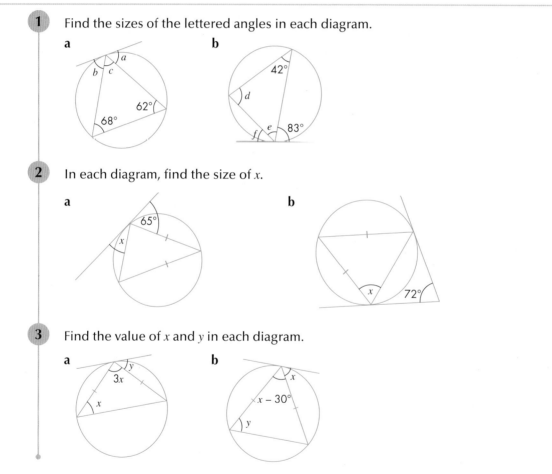

2 In each diagram, find the size of x.

a   **b**

3 Find the value of x and y in each diagram.

a   **b**

4 ATB is a tangent to the circle. O is the centre, where marked. Find the value of each lettered angle.

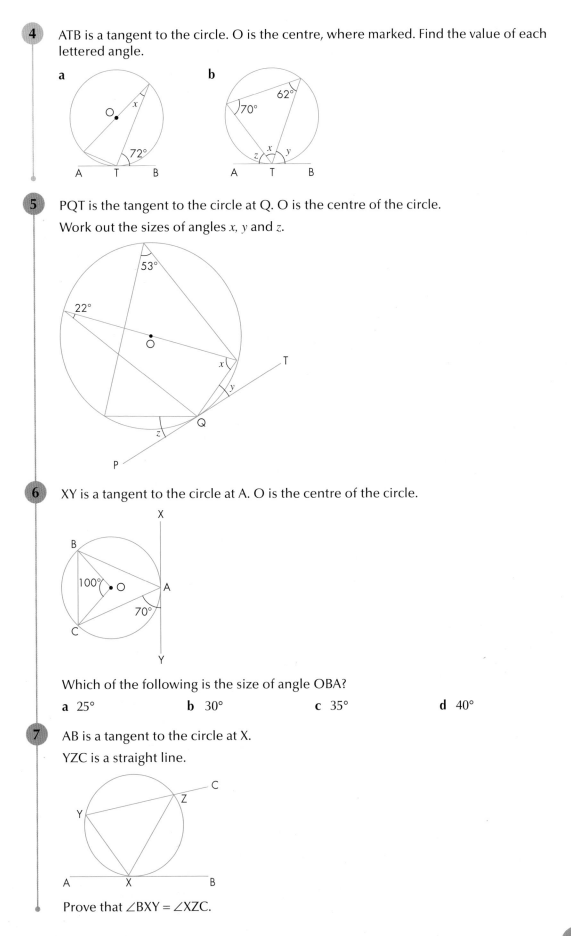

a

b

5 PQT is the tangent to the circle at Q. O is the centre of the circle.

Work out the sizes of angles x, y and z.

6 XY is a tangent to the circle at A. O is the centre of the circle.

Which of the following is the size of angle OBA?

a 25°　　　　　**b** 30°　　　　　**c** 35°　　　　　**d** 40°

7 AB is a tangent to the circle at X.

YZC is a straight line.

Prove that ∠BXY = ∠XZC.

21 Ratio and proportion and rates of change: Variation

21.1 Direct proportion

Homework 21A 🔲

For Questions **1** to **5**, first work out k, the constant of proportionality, and then the formula connecting the variables.

1 T is directly proportional to M. If $T = 30$ when $M = 5$, work out the value of:

    a T when $M = 4$ **b** M when $T = 75$.

2 W is directly proportional to F. If $W = 54$ when $F = 3$, work out the value of:

    a W when $F = 4$ **b** F when $W = 90$.

3 P is directly proportional to A. If $P = 50$ when $A = 2$, work out the value of:

    a P when $A = 5$ **b** A when $P = 150$.

4 A is directly proportional to t. If $A = 45$ when $t = 5$, work out the value of:

    a A when $t = 8$ **b** t when $A = 18$.

5 Q varies directly with P. If $Q = 200$ when $P = 5$, work out the value of:

    a Q when $P = 3$ **b** P when $Q = 300$.

6 The distance a train travels is directly proportional to the time taken for the journey. The train travels 135 miles in 3 hours.

    a What distance will the train travel in 4 hours?

    b How much time will it take for the train to travel 315 miles?

7 The cost of petrol is directly proportional to the amount put in a tank. When 40 litres is put in a tank, the cost is £32.00.

    a How much will it cost to put 30 litres in a tank?

    b How many litres can be put in a tank for £38.40?

    c A tank holds 60 litres when full. Petrol is put into the tank until it is full. The petrol costs £25.

    How much petrol was in the tank before it was filled up?

8 The number of people who can meet safely in a room is directly proportional to the area of the room. A room with an area of 200 m² is safe for 50 people.

   a How many people can safely meet in a room of area 152 m²?

   b A committee has 24 members. What is the smallest room area in which they could safely meet?

   c An extension is to be built on to a room which is safe for 50 people so that it can accommodate another 20 people.

     The cost of extending is estimated as £160 per square metre.

     How much is the estimate for the extension?

9 A man lays 36 paving stones in 3 hours.

   a Working at the same rate, how long would he take to lay 45 paving stones?

   b He works for 7 hours each day. He has 320 stones to lay.

     He employs another worker who can lay 10 stones each hour.

     Will they be able to complete the work in 2 days?

10 You are given that x varies directly with y.
The graph shows the relationship between x and y.

   Find the value of: **a** x when $y = 42$ **b** y when $x = 30$.

Homework 21B 🖩

For questions **1** to **5**, first work out k, the constant of proportionality, and then the formula connecting the variables.

1 T is directly proportional to x^2. If $T = 40$ when $x = 2$, work out the value of:

   a T when $x = 5$ **b** x when $T = 400$.

2 W is directly proportional to M^2. If $W = 10$ when $M = 5$, work out the value of:

   a W when $M = 4$ **b** M when $W = 64$.

3 A is directly proportional to r^2. If $A = 96$ when $r = 4$, work out the value of:

   a A when $r = 5$ **b** r when $A = 12$.

4 E varies directly with $\sqrt{C}$. If $E = 60$ when $C = 36$, work out the value of:

   a E when $C = 49$ **b** C when $E = 160$.

5 X is directly proportional to $\sqrt{Y}$. If $X = 80$ when $Y = 16$, work out the value of:

   a X when $Y = 100$ **b** Y when $X = 48$.

6 y is directly proportional to $\sqrt[3]{x}$. If $y = 4$ when $x = 8$, find the value of:

   a y when $x = 1$ **b** x when $y = 250$.

7 An artist is painting pictures.

The amount of time taken to complete a picture is directly proportional to the square of the width of the picture.

A picture that is 30 cm wide takes 20 hours to complete.

A buyer wants a picture that is 50 cm wide within 10 days.

If the artist paints for 6 hours each day, can he complete the picture on time?

8 Match each proportion statement with the correct sketch graph.

a $y \propto x^2$ **b** $y \propto x$ **c** $y \propto \sqrt{x}$

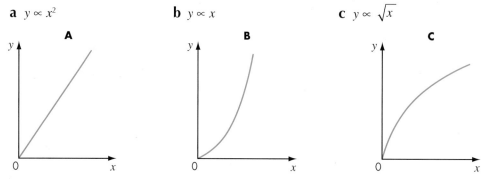

9 Match each table of values to one of the graphs in question **8**.

a

x	1	4	9
y	2	4	6

b

x	1	2	3
y	4	8	12

21.2 Inverse proportion

Homework 21C

For questions **1** to **7**, first find the equation connecting the variables.

1 T is inversely proportional to m. If $T = 7$ when $m = 4$, find the value of:

    a T when $m = 5$ **b** m when $T = 56$.

2 W is inversely proportional to x. If $W = 6$ when $x = 15$, find the value of:

    a W when $x = 3$ **b** x when $W = 10$.

3 M varies inversely with t^2. If $M = 10$ when $t = 2$, find the value of:

    a M when $t = 4$ **b** t when $M = 160$.

4 C is inversely proportional to f^2. If $C = 20$ when $f = 3$, find the value of:

    a C when $f = 5$ **b** f when $C = 720$.

5 W is inversely proportional to $\sqrt{T}$. If $W = 8$ when $T = 36$, find the value of:

    a W when $T = 25$ **b** T when $W = 0.75$.

6 H varies inversely with $\sqrt{g}$. If $H = 20$ when $g = 16$, find the value of:

    a H when $g = 1.25$ **b** g when $H = 40$.

7 y is inversely proportional to the cube of x. If $y = 10$ when $x = 1$, find the value of:

    a y when $x = 2$ **b** x when $y = 270$.

8 The brightness of the light from a bulb decreases inversely with the square of the distance from the bulb. The brightness is 5 candle power at a distance of 10 m. What is the brightness at a distance of 5 m?

9 In the table, y is inversely proportional to x.

x	2	4	16
y	8		

Copy and complete the table.

10 The density of a series of spheres with the same mass is inversely proportional to the cube of the radius.
The graph shows the relationship between the density (d) and the radius (r).

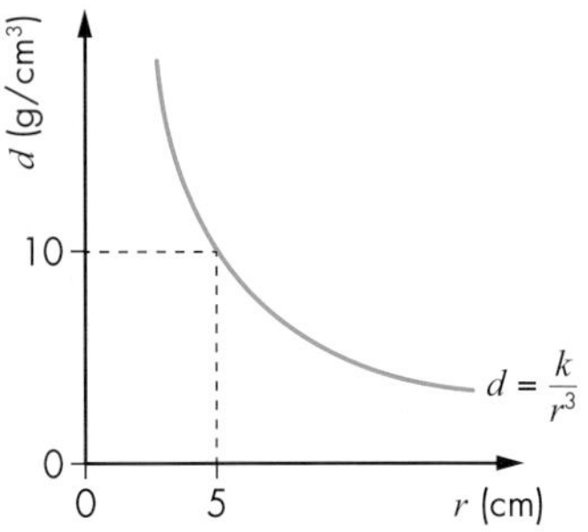

a What would be the density of a sphere with a radius of 10 cm?

b What would be the radius of a sphere with density 80 g/cm³?

11 y is inversely proportional to the square of x. When $x = 4$, y is 12.

a Find an expression for y in terms of x.

b Find the value of:

   i y when $x = 6$ ii x when $y = 36$.

12 The time taken to build an extension is inversely proportional to the number of workers.

It takes 2 workers 7 days to complete an extension.

a 3 workers start an extension on Monday morning.

   Will they complete it by Friday?

   Show your working.

b Give a reason why the time taken might not be inversely proportional to the number of workers when the number of workers is very large.

13 a Which statement does the graph represent?

   A: $y \propto x$ **B:** $y \propto \dfrac{1}{x}$ **C:** $y \propto \sqrt{x}$

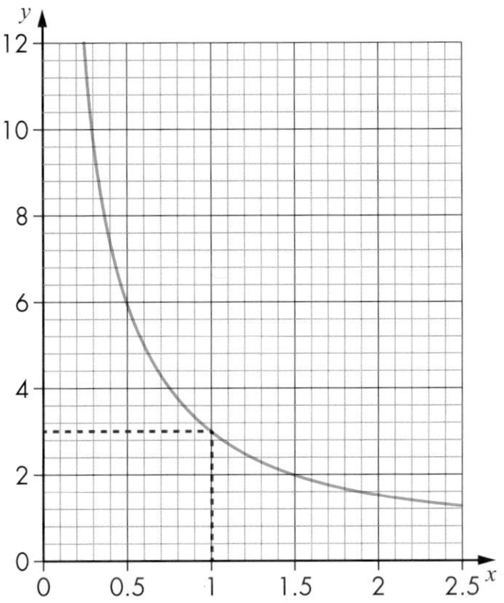

b Use your answer to find an equation for the graph.

22 Geometry and measures: Triangles

22.1 Further 2D problems

Homework 22A 🖩

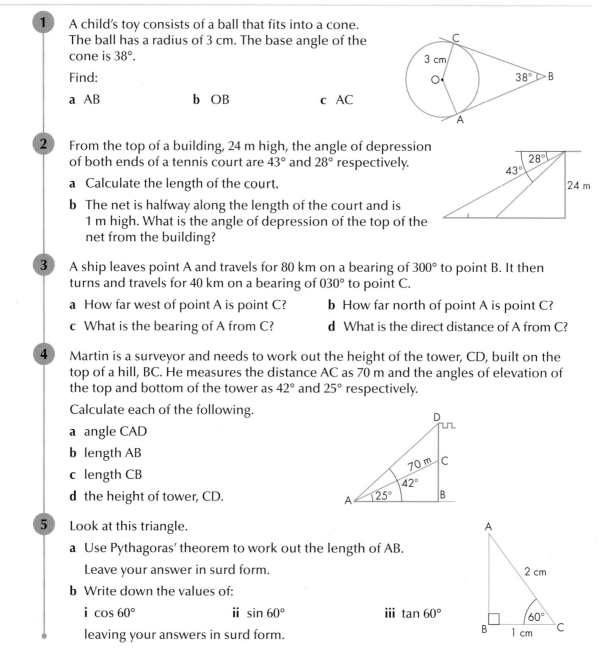

1 A child's toy consists of a ball that fits into a cone. The ball has a radius of 3 cm. The base angle of the cone is 38°.

Find:

  a AB   **b** OB   **c** AC

2 From the top of a building, 24 m high, the angle of depression of both ends of a tennis court are 43° and 28° respectively.

  a Calculate the length of the court.

  b The net is halfway along the length of the court and is 1 m high. What is the angle of depression of the top of the net from the building?

3 A ship leaves point A and travels for 80 km on a bearing of 300° to point B. It then turns and travels for 40 km on a bearing of 030° to point C.

  a How far west of point A is point C?   **b** How far north of point A is point C?

  c What is the bearing of A from C?   **d** What is the direct distance of A from C?

4 Martin is a surveyor and needs to work out the height of the tower, CD, built on the top of a hill, BC. He measures the distance AC as 70 m and the angles of elevation of the top and bottom of the tower as 42° and 25° respectively.

Calculate each of the following.

  a angle CAD

  b length AB

  c length CB

  d the height of tower, CD.

5 Look at this triangle.

  a Use Pythagoras' theorem to work out the length of AB. Leave your answer in surd form.

  b Write down the values of:

    i cos 60°   **ii** sin 60°   **iii** tan 60°

    leaving your answers in surd form.

6 In the diagram, triangle ACD is right-angled and triangle ABC is isosceles.

Calculate the size of angle ABC.

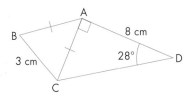

7 A regular pentagon is inscribed in a circle of radius 5 cm. Calculate the length of one of its sides.

22.2 Further 3D problems

Homework 22B

1 A TV mast XY is 3 km due west of village A.

Village B is 2 km due south of village A.

The angle of elevation from B to the top of the mast is 6°.

Show how a surveyor can use this information to calculate the height of the mast in metres.

2 The base of this pyramid, ABCD, is a square with side length 16 cm. The length of each sloping edge is 25 cm. The apex, V, is over the centre of the square base.

Calculate:

a the size of angle VAC

b the height of the pyramid

c the volume of the pyramid

d the size of the angle between the face VAD and the base ABCD.

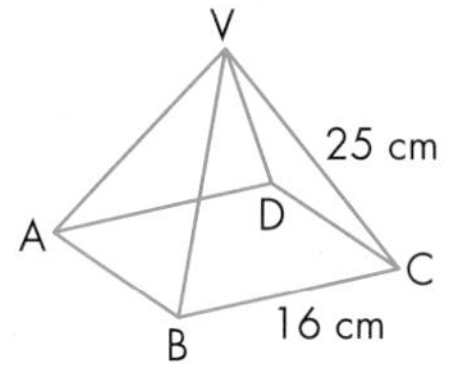

3 In the diagram, M is the midpoint of GH.

Find the size of these angles.

a angle AGE

b angle BMA

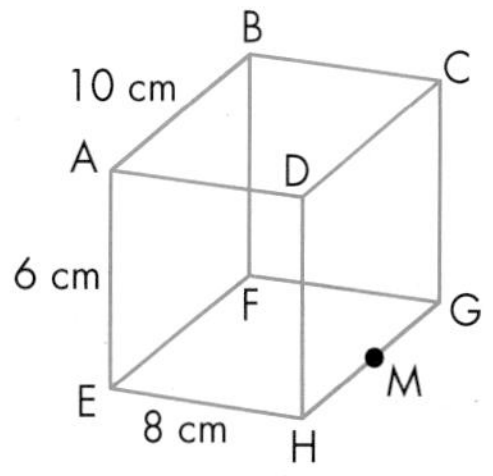

4 M is the midpoint of AB in this wedge. Calculate the size of each of these.

a length CD   **b** angle CAD

c angle CAE   **d** length DM

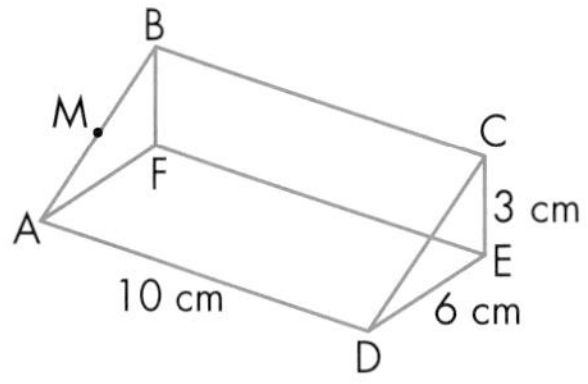

5 The diagram shows a tetrahedron VPQR on top of a prism FGHRQP.
The cross-section of the prism, PQR, is an equilateral triangle of side 8 cm.

VP = VQ = VR = 10 cm

PF = QG = RH = 15 cm

M is the midpoint of QR.

a i Use triangle PQR to find the length of PM.

   ii Use triangle VQR to find the length of VM.

b Find the size of angle VPM.

c Find the height of V above the base FGH. Give your answer to an appropriate degree of accuracy.

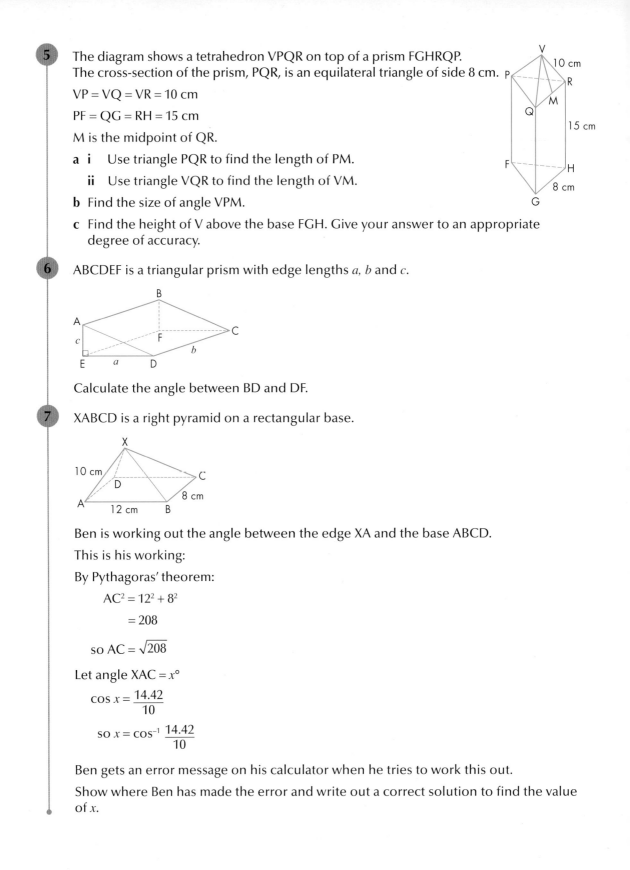

6 ABCDEF is a triangular prism with edge lengths a, b and c.

Calculate the angle between BD and DF.

7 XABCD is a right pyramid on a rectangular base.

Ben is working out the angle between the edge XA and the base ABCD.

This is his working:

By Pythagoras' theorem:

$$AC^2 = 12^2 + 8^2$$

$$= 208$$

so $AC = \sqrt{208}$

Let angle XAC = $x°$

$$\cos x = \frac{14.42}{10}$$

so $x = \cos^{-1} \frac{14.42}{10}$

Ben gets an error message on his calculator when he tries to work this out.

Show where Ben has made the error and write out a correct solution to find the value of x.

22.3 Trigonometric ratios of angles between 0° and 360°

Homework 22C 🖩

1 State the two angles between 0° and 360° for each of these sine values.

    a 0.4   **b** 0.45   **c** 0.65   **d** 0.27

    e 0.453   **f** −0.4   **g** −0.15   **h** −0.52

2 Solve the equation $2 \sin x = 1$ giving all answers between 0° and 360°.

3 Given that $\sin 40° = 0.643$, write down these values.

    a sin 140°   **b** sin 320°   **c** sin 400°   **d** sin 580°

4 Solve the equation $3 \sin x = -2$ giving all answers between 0° and 360°.

5 Which of these values is the odd one out? Give a reason for your answer.

    sin 36°   sin 78°   sin 119°   sin 320°

6 State the two angles between 0° and 360° for each of these cosine values.

    a 0.7   **b** 0.38   **c** 0.617   **d** 0.376

    e 0.085   **f** −0.6   **g** −0.45   **h** −0.223

7 Solve the equation $3 \cos x = -1$ giving all answers between 0° and 360°.

8 Give that $\cos 50° = 0.643$, write down these values.

    a cos 130°   **b** cos 310°   **c** cos 410°   **d** cos 590°

9 Solve the equation $6 \cos x = -1$ giving all answers between 0° and 360°.

10 Which of these values is the odd one out. Give a reason for your answer.

    cos 68°   cos 112°   cos 248°   cos 338°

Homework 22D 🖩

1 Write down the sine of each of these angles.

    a 27°   **b** 153°   **c** 207°   **d** 333°

2 Write down the cosine of each of these angles.

    a 69°   **b** 111°   **c** 249°   **d** 291°

3 What do you notice about the answers to questions **1** and **2**?

4 In each case, find four values of x between 0° and 360° such that:

    a $\sin x = \pm 0.4$ **b** $\cos x = \pm 0.5$.

5 Write down the value of each of the following, correct to 3 significant figures.

    a sin 40° + cos 60°   **b** cos 130° − sin 130°   **c** sin 145° + cos 226°

    d sin 115° + cos 250°   **e** sin 116° − sin 220°   **f** cos 125° + sin 179°

6 Suppose the cosine key on your calculator is broken but the sine key is working. Show how you could calculate these.

    a cos 25°   **b** cos 130°

7 Find the solution for each equation.

    a $\sin x + 1 = 2$ for $0° \leqslant x \leqslant 360°$ **b** $2 + 3 \cos x = 1$ for $0° \leqslant x \leqslant 360°$

8 Find two values of x between 0° and 360° such that $\sin x = \cos 320°$.

9 Find a solution for each equation.

    a $\sin (x + 30°) = 0.6$ **b** $\cos (3x) = -0.866$

Homework 22E

1 State the angles between 0° and 360° for each of these tangent values.

a 0.528	**b** 0.8	**c** 1.35	**d** 3.24
e −2.55	**f** −0.158	**g** −0.786	**h** −1.999

2 Given that tan 64° = 2.05, write down the tangent values of these angles.

    a 116°   **b** 296°   **c** 424°   **d** 604°

3 Which of these values is the odd one out? Give a reason for your answer.

    tan 135°   tan 315°   tan 495°   tan 585°

4 Write down two angles between −180° and 180° that have a tangent value of −1.191.

22.4 Solving any triangle

Homework 22F

1 Work out the length of x in each triangle.

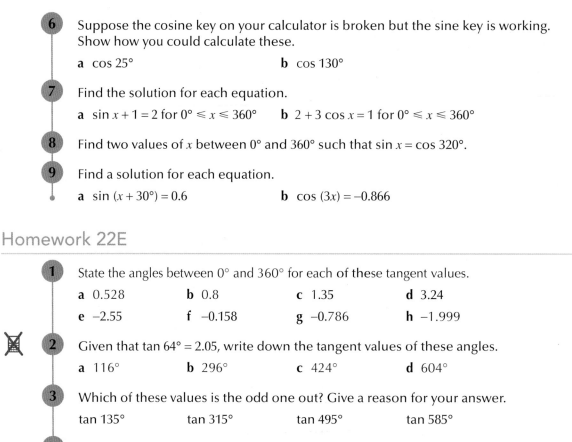

a (triangle with B 87°, A 62°, side AC = 5 m, x opposite)

b (triangle with angle 112° and 38° at C, x, side = 15 cm)

c (triangle with B, x, 9 m, A 63°, side AC = 5 m)

d (triangle with B, 17 cm, x, C 40°, 16 cm)

2 In triangle ABC, angle A is 40°, side AB is 10 cm and side BC is 7 cm. Work out the two possible values of angle C.

3 In triangle ABC, angle A is 58°, side AB is 20 cm and side BC is 18 cm. Work out the two possible lengths of side AC.

4 To calculate the length of a submarine, Mervyn stood on a cliff 60 m high and made some measurements (see diagram).

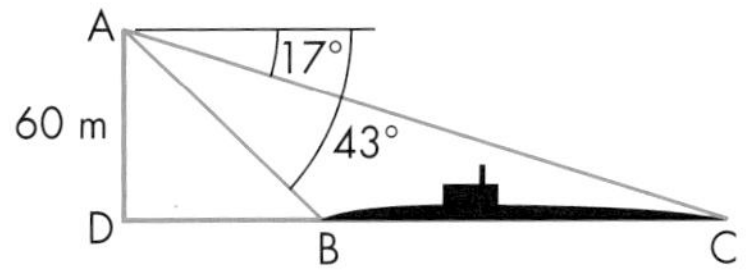

a Calculate the size of angle DAB.

b Use trigonometry to calculate the length AB.

c Use the sine rule to work out the length BC.

5 Use the information on this sketch to help the land surveyor calculate the width, w, of the river.

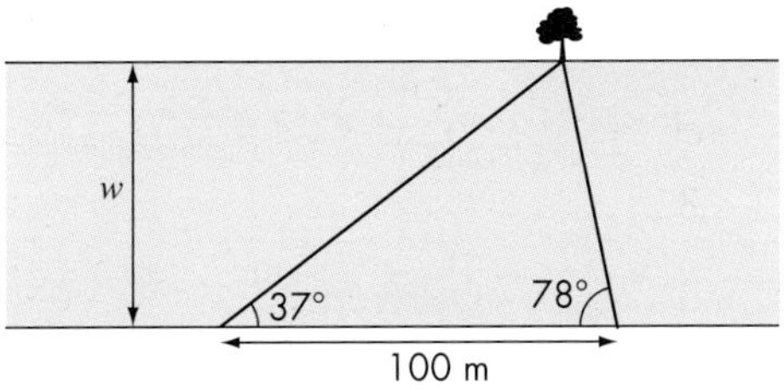

6 A surveyor needs to find the height of a chimney. She measures the angle of elevation as 28°. She then walks 30 m towards the chimney and measures the angle of elevation from this point as 37°. What is the height of the chimney?

7 The diagram shows ship S and two lighthouses, A and B.

A is due west of B and the two lighthouses are 15 km apart.

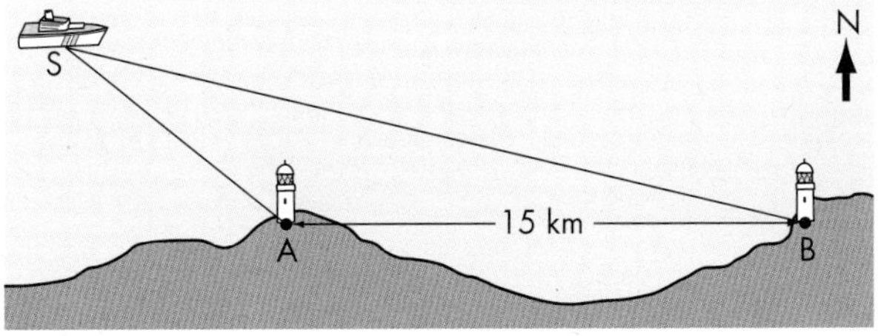

The bearing of the ship from lighthouse A is 330° and the bearing of the ship from lighthouse B is 290°.

How far is the ship from lighthouse B?

8 Triangle ABC has an obtuse angle at A. Calculate the size of angle BAC.

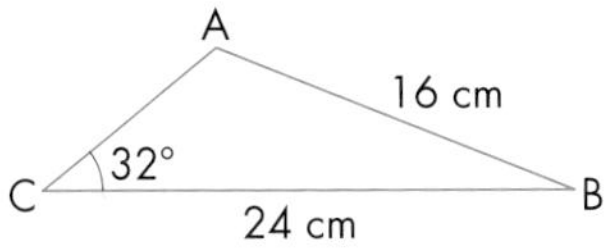

1 Calculate the length of x in each triangle.

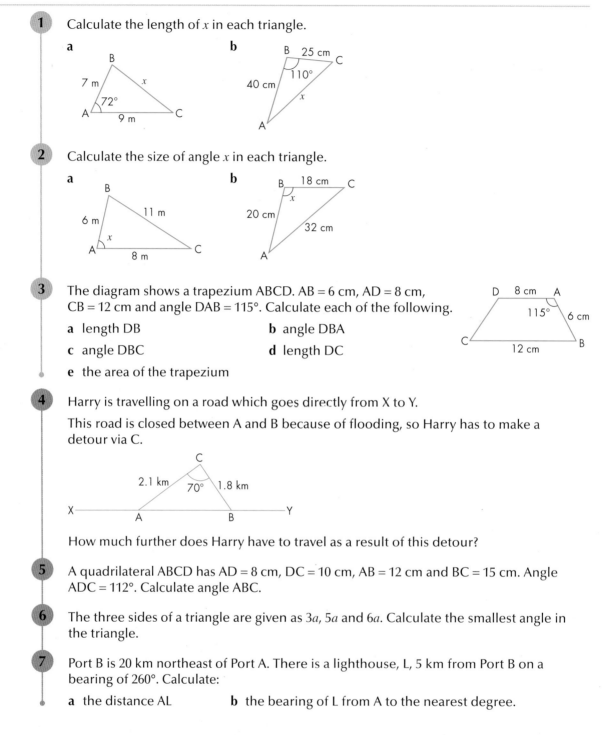

a

b

2 Calculate the size of angle x in each triangle.

a

b

3 The diagram shows a trapezium ABCD. AB = 6 cm, AD = 8 cm, CB = 12 cm and angle DAB = 115°. Calculate each of the following.

a length DB

b angle DBA

c angle DBC

d length DC

e the area of the trapezium

4 Harry is travelling on a road which goes directly from X to Y.

This road is closed between A and B because of flooding, so Harry has to make a detour via C.

How much further does Harry have to travel as a result of this detour?

5 A quadrilateral ABCD has AD = 8 cm, DC = 10 cm, AB = 12 cm and BC = 15 cm. Angle ADC = 112°. Calculate angle ABC.

6 The three sides of a triangle are given as $3a$, $5a$ and $6a$. Calculate the smallest angle in the triangle.

7 Port B is 20 km northeast of Port A. There is a lighthouse, L, 5 km from Port B on a bearing of 260°. Calculate:

a the distance AL

b the bearing of L from A to the nearest degree.

8 Calculate the size of the smallest angle in triangle XYZ.

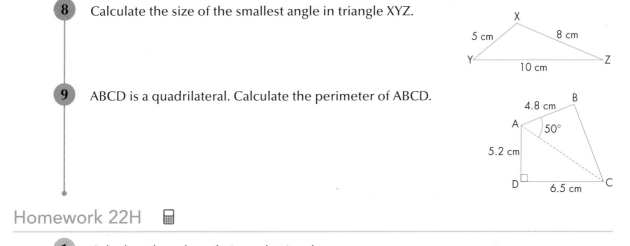

9 ABCD is a quadrilateral. Calculate the perimeter of ABCD.

Homework 22H

1 Calculate the value of x in each triangle.

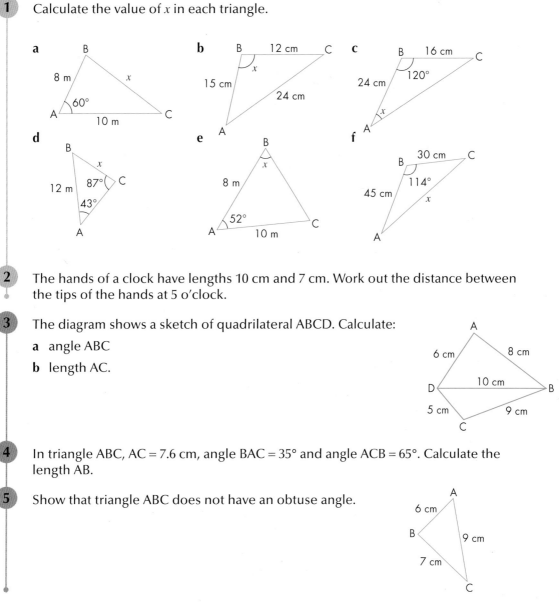

2 The hands of a clock have lengths 10 cm and 7 cm. Work out the distance between the tips of the hands at 5 o'clock.

3 The diagram shows a sketch of quadrilateral ABCD. Calculate:

a angle ABC

b length AC.

4 In triangle ABC, AC = 7.6 cm, angle BAC = 35° and angle ACB = 65°. Calculate the length AB.

5 Show that triangle ABC does not have an obtuse angle.

22.5 Using sine to calculate the area of a triangle

Homework 22I

1 Work out the area of these triangles.

 a Triangle ABC where BC = 8 cm, AC = 10 cm and angle ACB = 69°.

 b Triangle PQR where angle QPR = 112°, PR = 3 cm and PQ = 7 cm.

2 The area of triangle ABC is 27 cm². If BC = 12 cm and angle BCA = 98°, calculate AC.

3 The area of triangle LMN is 85 cm², LM = 10 cm and MN = 25 cm. Calculate these angles.

 a LMN   b MNL

4 In a quadrilateral ABCD, DC = 3 cm, BD = 8 cm, angle BAD = 43°, angle ABD = 52° and angle BDC = 72°. Calculate the area of the quadrilateral.

5 A signwriter wants to paint a board in the shape of a triangle with sides 30 cm, 40 cm and 60 cm. Work out the area of the board.

6 In triangle ABC, ∠BAC = 32°, AC = 10 cm and BC = 6 cm. Angle B is obtuse. Calculate the area of the triangle ABC.

7 The diagram shows a sketch of some land that a farmer plans to use as an orchard.

 a Calculate the area of the orchard. Give your answer to an appropriate degree of accuracy.

 b The trees the farmer wants to plant each need 5 m² of space. How many trees can he plant in the orchard?

8 ABCD is a parallelogram. AB = a and BC = b. Angle ABC = θ.

 Prove that the area of the parallelogram is given by the formula:

   area = $ab \sin \theta$.

9 ABCD is a quadrilateral.

 Work out the area of the quadrilateral.

 Give your answer to an appropriate degree of accuracy.

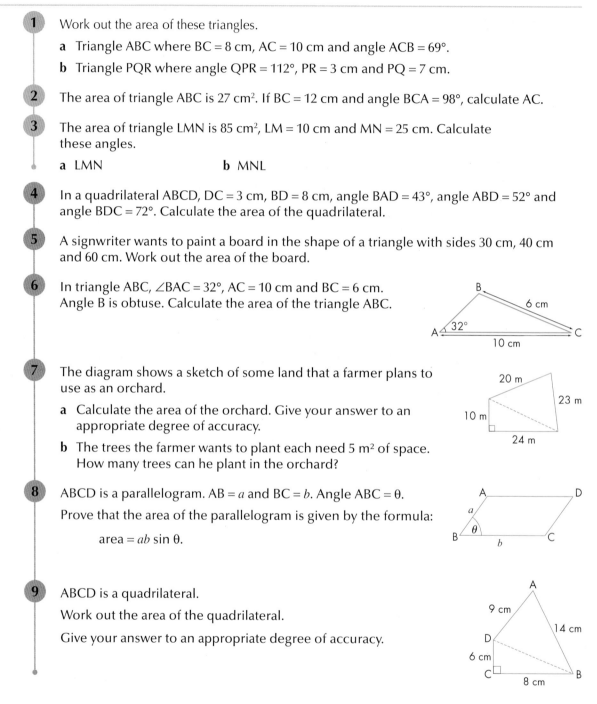

23 Algebra: Graphs

23.1 Distance–time graphs

Homework 23A

1 This distance–time graph illustrates Joe's car journey to meet his girlfriend. He set off from home at 9.00 am and stopped on the way for a break.

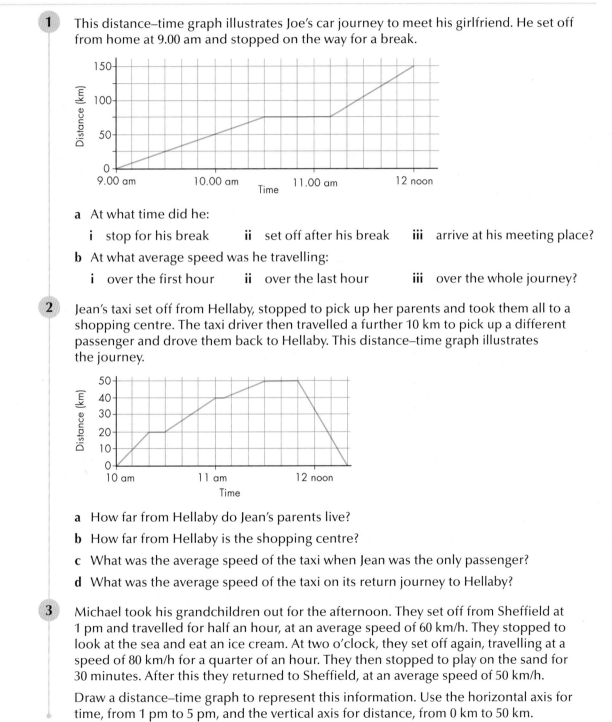

a At what time did he:

 i stop for his break **ii** set off after his break **iii** arrive at his meeting place?

b At what average speed was he travelling:

 i over the first hour **ii** over the last hour **iii** over the whole journey?

2 Jean's taxi set off from Hellaby, stopped to pick up her parents and took them all to a shopping centre. The taxi driver then travelled a further 10 km to pick up a different passenger and drove them back to Hellaby. This distance–time graph illustrates the journey.

a How far from Hellaby do Jean's parents live?

b How far from Hellaby is the shopping centre?

c What was the average speed of the taxi when Jean was the only passenger?

d What was the average speed of the taxi on its return journey to Hellaby?

3 Michael took his grandchildren out for the afternoon. They set off from Sheffield at 1 pm and travelled for half an hour, at an average speed of 60 km/h. They stopped to look at the sea and eat an ice cream. At two o'clock, they set off again, travelling at a speed of 80 km/h for a quarter of an hour. They then stopped to play on the sand for 30 minutes. After this they returned to Sheffield, at an average speed of 50 km/h.

Draw a distance–time graph to represent this information. Use the horizontal axis for time, from 1 pm to 5 pm, and the vertical axis for distance, from 0 km to 50 km.

4 A runner sets off at 8 am from point P to jog along a trail at a steady pace of 12 km/h.

One hour later, a cyclist sets off from P on the same trial, at a steady pace of 24 km/h. After 30 minutes, the cyclist gets a puncture that takes her 30 minutes to fix. She then sets off at a steady pace of 24 km/h.

At what time did the cyclist catch up with the runner?

> **Hints and tips** Drawing a distance–time graph is a straightforward method of answering this question. Remember that the cyclist doesn't start until 9 am.

5 Calculate the average speed of each journey.

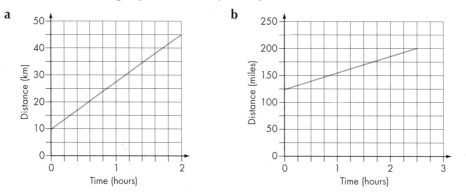

a

b

6 This graph shows a car's journey from London to Brighton and back again. The car leaves at 8 am and returns at 3 pm.

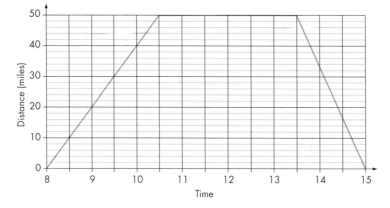

a For how long does the car stop in Brighton?

b Was the car travelling faster from London to Brighton or on the return journey from Brighton to London? Describe how you can tell this from the graph.

Homework 23B

1 Draw a graph of the depth of water in each of these containers as it is filled steadily.

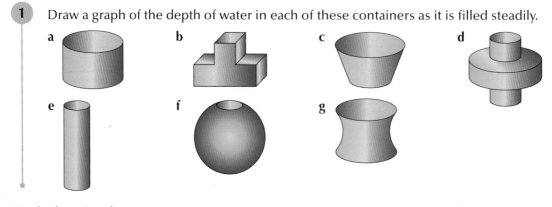

a **b** **c** **d**

e **f** **g**

2 The graph shows the depth of water in Melvin's bath from the time he started running the water to the time that the bath was empty again.

    a Explain what you think is happening for each part of the graph from **a** to **g**.

    b Draw the cross-section of the bath.

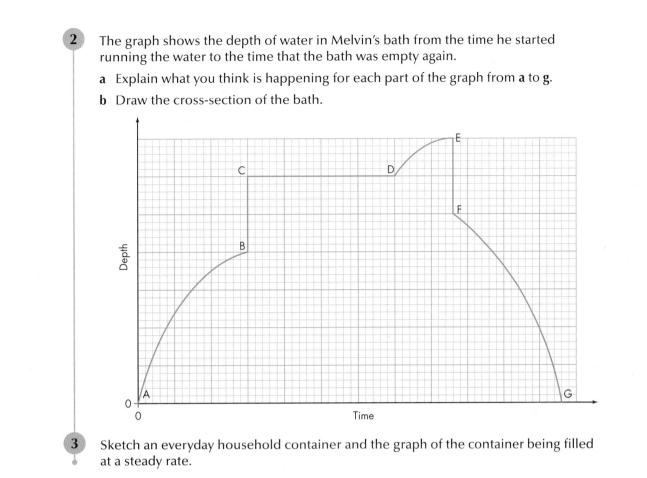

3 Sketch an everyday household container and the graph of the container being filled at a steady rate.

23.2 Velocity–time graphs

Homework 23C

1 **a** Work out the speed for the first part of the journey shown in this distance–time graph.

    b For the second part of the journey, was the vehicle travelling faster or more slowly? Give a reason for your answer.

    c Work out the average speed for the whole journey.

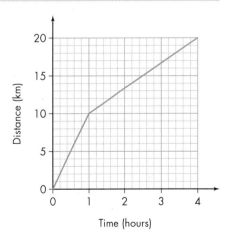

2 A car increases its speed at a steady rate from rest (0 m/s) to 20 m/s in 10 seconds.

It then travels at a steady speed for 30 seconds before increasing its speed at a steady rate to 30 m/s over 10 seconds. It then slows down to rest over a further 20 seconds, decreasing speed at a steady rate.

    a Draw a graph to represent this information.

    b Use your graph to work out the total distance the car travels.

3 This graph represents a train journey.

    a Work out speed of the train between A to B?

    b For how long was the train stationary at B?

    c A second train starts its journey from C at 11 am and travels to A, without stopping, at an average speed of 60 mph. Copy the graph and draw a line to represent the journey of the second train.

    d How far from A were the trains when they passed each other?

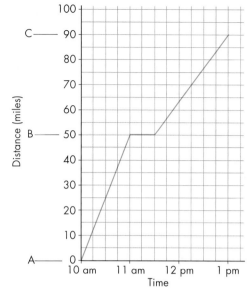

Homework 23D

1 The diagram shows the velocity of a car over 10 seconds.

Calculate the acceleration:

    a over the first 2 seconds

    b after 6 seconds.

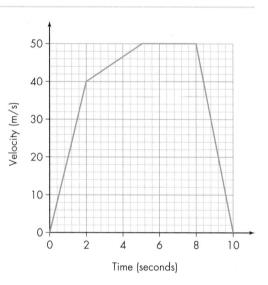

2 The graph shows the journey of a car travelling on a busy road.

    a Find the initial acceleration of the car.

    b Find the final deceleration of the car.

    c For how long was the car not accelerating or decelerating?

    d Find the distance travelled whilst the car was travelling at a steady speed.

    e Find the total distance travelled.

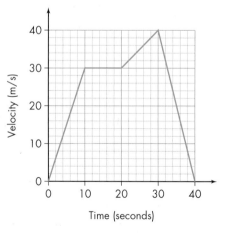

3 The graph shows the journey of a train travelling between two towns.

a Work out the acceleration or deceleration for each section of the graph.

b Find the total distance travelled.

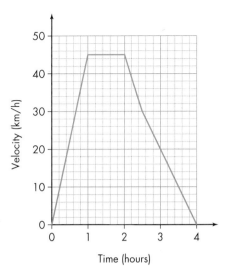

4 The diagram shows the velocity–time graph for a train journey between two stations.

a Find the acceleration over the first 10 seconds.

b Find the deceleration over the last 20 seconds.

c Find the distance between the two stations.

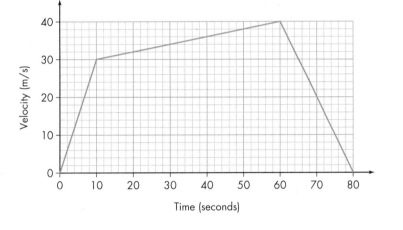

5 The graph shows the velocity of a car between two junctions.

a Work out the acceleration in the first 15 seconds, in terms of v.

b The distance travelled in the first 15 seconds is 375 metres. Work out the total distance travelled between the junctions.

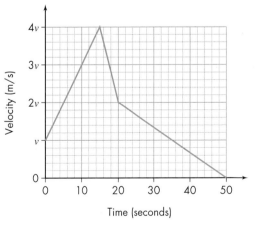

23.3 Estimating the area under a curve

1 For each of the velocity–time graphs, estimate the distance travelled and state whether your estimate is an under-estimate or over-estimate.

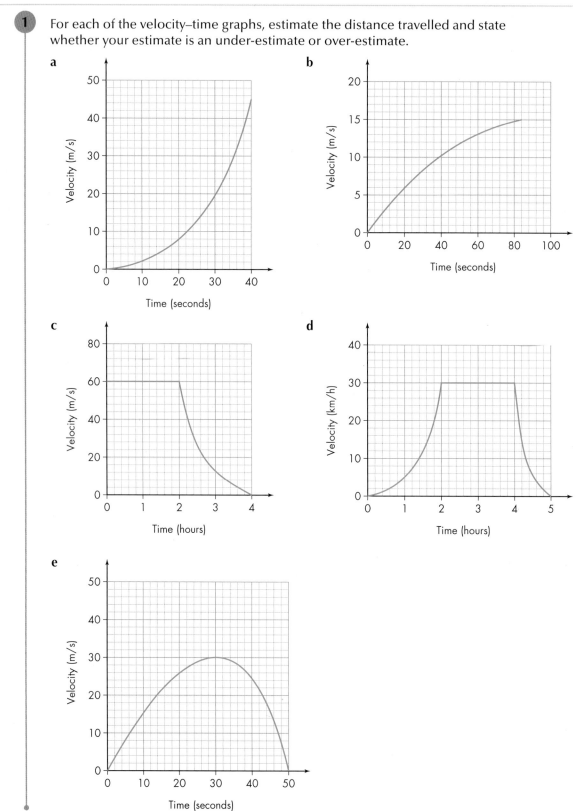

2　**a** For this velocity–time graph, find:

　　i the initial velcity

　　ii the maximum velocity.

　b Find an estimate for the total distance travelled. State whether it is an underestimate or overestimate.

　c How could you make your answer to part **b** more accurate?

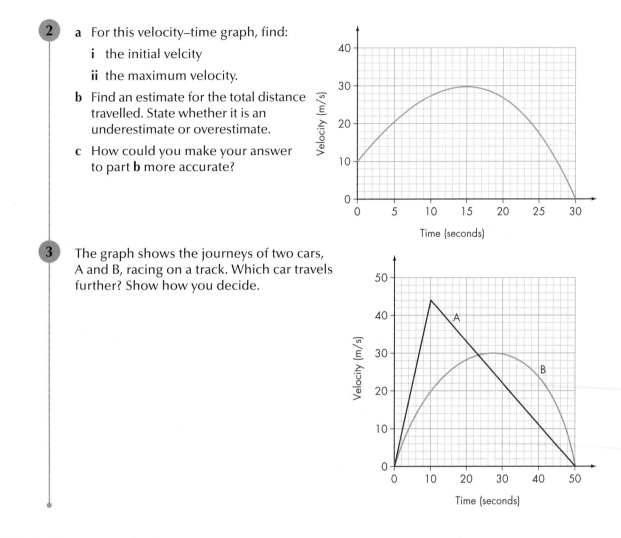

3　The graph shows the journeys of two cars, A and B, racing on a track. Which car travels further? Show how you decide.

23.4 Rates of change

Homework 23F

1　The graph shows the height of a stone as it is dropped from the top of a building.

　a Draw a tangent at the point where $t = 3$.

　b Use your tangent to estimate the velocity of the stone after 3 seconds.

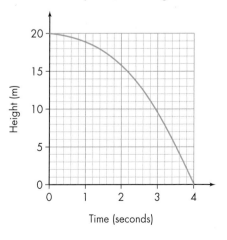

2 Look at this distance–time graph.

a Estimate the velocity when:

   i $t = 2$ **ii** $t = 4.5$.

b At what times is the velocity zero?

c Estimate average velocity from $t = 0$ to $t = 1$.

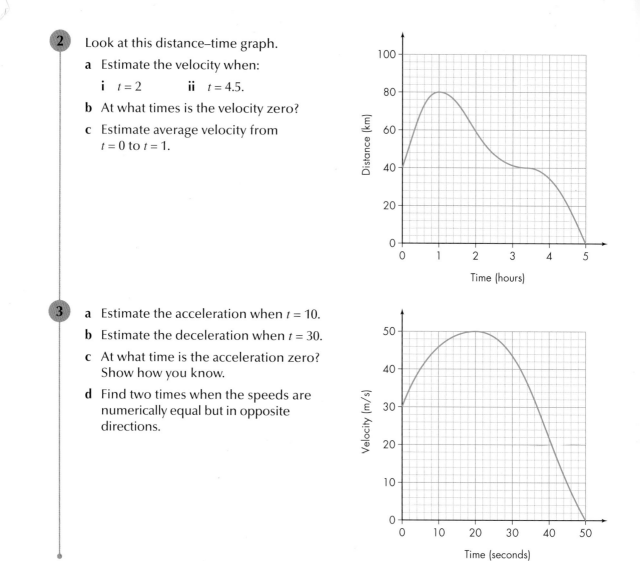

3 a Estimate the acceleration when $t = 10$.

b Estimate the deceleration when $t = 30$.

c At what time is the acceleration zero? Show how you know.

d Find two times when the speeds are numerically equal but in opposite directions.

23.5 Equation of a circle

Homework 23G

1 State the radius of each circle. Give your answers as simplified surds, where appropriate.

  a $x^2 + y^2 = 25$ **b** $x^2 + y^2 = 8$ **c** $x^2 + y^2 = 289$ **d** $x^2 + y^2 = 529$

2 State the diameter of each circle. Give your answers as simplified surds, where appropriate.

  a $x^2 + y^2 = 125$ **b** $x^2 + y^2 = 3136$ **c** $x^2 + y^2 = 900$ **d** $x^2 + y^2 = \frac{4}{25}$

3 A circle has equation $x^2 + y^2 = 144$. Determine whether each point lies inside the circle, outside the circle or on the circumference of the circle.

  a (3, 4)   **b** (8, 8)   **c** (–7, 10)   **d** (12, 0)

> **Hints and tips** Use Pythagoras' theorem to work out how far each point is from the origin and compare it with the radius.

4 (12, 5) and (0, –13) are coordinates of points on the circumference of the circle with equation $x^2 + y^2 = 169$, where both x and y are integers.

    a State three other points on the circumference of $x^2 + y^2 = 169$ where the coordinates are integers.

    b How many pairs of coordinates on the circumference of $x^2 + y^2 = 169$ have both values as integers?

5 Point A(3, 5) lies on the circumference of the circle $x^2 + y^2 = 34$.

    a Find the gradient of the line segment joining A to the origin (0, 0).

    b Find the gradient of the tangent to the circle at A.

    c Find the equation of the tangent to the circle at A in the form $y = mx + c$.

6 Find the equation of the tangent in the form $y = mx + c$ for these situations.

    a Circle $x^2 + y^2 = 29$, tangent at (2, –5)

    b Circle $x^2 + y^2 = 25$, tangent at (–3, –4)

7 Find the equations of both tangents to the circle $x^2 + y^2 = 90$ with a gradient of 3. Give your answers in the form $y = mx + c$.

8 A circle has the equation $x^2 + y^2 = 72$.

    Find the equations of the tangents to the circle that are perpendicular to $y = -x - 4$.

9 A circle centred at the origin has a tangent with equation $y = \frac{2}{9}x + c$ at the point (–2, 9).

    a Find the value of c.

    b Find the equation of the circle.

23.6 Other graphs

Homework 23H

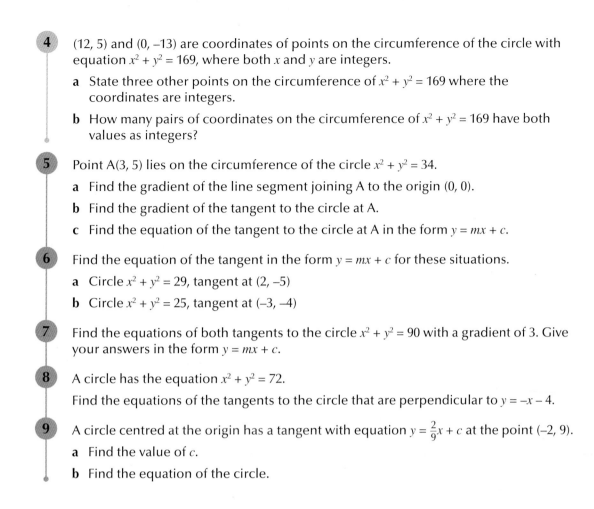

1   **a** Copy and complete the table and draw the graph of $y = x^3 + 1$ for $-3 \le x \le 3$.

x	–3	–2	–1	0	1	2	3
y	–26		1				28

    b Use your graph to find the y-value when $x = 1.2$.

2   **a** Copy and complete the table and draw the graph of $y = x^3 + 2x$ for $-2 \le x \le 3$.

x	–2	–1	0	1	2	3
y	–12		0		12	

    b Use your graph to find the y-value when $x = 2.5$.

3   **a** Copy and complete the table and draw the graph of $y = \frac{12}{x}$ for $-12 \le x \le 12$.

x	–12	–6	–4	–2	–1	–0.5	0.5	1	2	4	6	12
y	–1			–6					6			1

    b Use your graph to find the:

      i y-value when $x = 1.5$ **ii** x-value when $y = 5.5$.

4 **a** Copy and complete the table and draw the graph of $y = \frac{50}{x}$ for $0 \le x \le 50$.

x	0.5	1	2	5	10	25	50
y							

 b On the same axes, draw the graph of $y = x + 30$.

 c Use your graph to find the x-values of any points where the graphs intersect.

5 **a** Copy and complete the table for $y = 2^x$ for values of x from -3 to $+4$. (Values are rounded to 1 dp.)

x	-3	-2	-1	0	1	2	3	4
y	0.1	0.3			2	4		

 b Plot the graph of $y = 2^x$ for $-3 \le x \le 4$. (Take y-axis from 0 to 20.)

 c Use your graph to estimate the value of y when $x = 2.5$.

 d Use your graph to estimate the value of x when $y = 0.75$.

6 A curve of the form $y = ab^x$ passes through the points $(0, 3)$ and $(2, 48)$. Work out the values of a and b.

23.7 Transformations of the graph $y = f(x)$

Homework 23I

You may use a graphical calculator or a graph drawing program to complete this exercise.

1 **a** Plot these graphs on the same axes.

  i $y = x^2$ **ii** $y = x^2 + 2$ **iii** $y = (x + 2)^2$

 b Describe the relationship between $y = x^2$ and each of the other graphs in parts **a**.

2 **a** Plot these graphs on the same axes.

  i $y = x^2$ **ii** $y = x^2 - 3$ **iii** $y = x^2 + 1$

 b Describe the relationship between $y = x^2$ and each of the other graphs in parts **a**.

3 **a** Plot these graphs on the same axes.

  i $y = x^2$ **ii** $y = (x + 4)^2$ **iii** $y = -x^2$ **iv** $y = 2 - x^2$

 b Describe the relationship between $y = x^2$ and each of the other graphs in parts **a**.

Homework 23J

1 **a** Plot these graphs on the same axes.

  i $y = \sin x$ **ii** $y = \sin x + 3$ **iii** $y = \sin (x + 30°)$

 b Describe the transformation that takes the graph in part **i** to each of the other graphs.

2 **a** Plot these graphs on the same axes.

  i $y = \sin x$ **ii** $y = \sin x + 2$ **iii** $y = \sin (x + 45°)$

 b Describe the transformation that takes the graph in part **i** to each of the other graphs.

3 **a** Plot these graphs on the same axes.

   i $y = \cos x$ **ii** $y = -\cos x$ **iii** $y = \cos x + 4$

   b Describe each relationship between the graph in part **i** and the graphs in parts **ii** and **iii**.

4 **a** Plot these graphs on the same axes.

   i $y = \cos x$ **ii** $y = \cos(x + 60°)$ **iii** $y = \cos x + 3$

   b Describe each relationship between the graph in part **i** and the graphs in parts **ii** and **iii**.

5 Show that the graphs of $y = \cos x$ and $y = \sin(x + 90°)$ are the same.

6 The graphs are all transformations of $y = x^2$. The coordinates of two points are marked on each graph. Use this information to work out the equation of each graph.

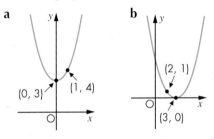

a   **b**

7 Below are the graphs of: $y = -\sin x$ and $y = \cos x$.

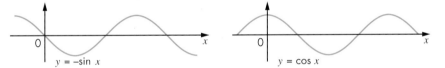

$y = -\sin x$ $y = \cos x$

   a Describe a series of transformations that would take $y = -\sin x$ to $y = \cos x$.

   b Which of these is equivalent to $y = -\sin x$?

   i $y = \sin(x + 180°)$ **ii** $y = \cos(x + 90°)$ **iii** $y = \sin \frac{x}{2}$

8 The graph shows the function $y = f(x)$. Use this to sketch these functions.

   a $y = f(x) - 2$ **b** $y = f(x - 2)$

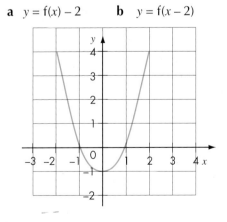

1 The table shows some values of the function $f(x) = (x - 2)^2 + 4$, where $-3 \leq x \leq 4$.

x	-3	-2	-1	0	1	2	3	4
$f(x)$	29	20	13	8	5	4	5	8

a Draw the graph of $y = f(x)$.

b On the same axes, draw the graph of $y = x^2$.

c Describe how the graph of $y = (x - 2)^2 + 4$ can be obtained from the graph $y = x^2$ by a transformation. State clearly what this transformation is.

2 Given that $f(x) = x^2$, write each equation in the form $y = f(x \pm a) \pm b$.

State the translation of the graph in each case.

a $y = x^2 - 4x + 6$ **b** $y = x^2 + 10x + 15$ **c** $y = x^2 - 20x + 90$

3 Sketch each of these graphs, using completing the square to find the graph transformation.

In each case, show the minimum or maximum point and the intersection with the y-axis.

a $y = x^2 + 4x + 2$ **b** $y = x^2 - 8x + 14$ **c** $y = -16 - x^2 + 8x$

24 Algebra: Algebraic fractions and functions

24.1 Algebraic fractions

Homework 24A

1 Simplify each of these.

a $\dfrac{2x}{3} + \dfrac{4x}{5}$

b $\dfrac{x+1}{3} + \dfrac{x+3}{2}$

c $\dfrac{2x-3}{2} + \dfrac{5x-1}{3}$

2 Simplify each of these.

a $\dfrac{3x}{4} - \dfrac{2x}{5}$

b $\dfrac{x+2}{2} - \dfrac{x+1}{5}$

c $\dfrac{4x-1}{2} - \dfrac{2x-4}{3}$

3 Solve these equations.

a $\dfrac{2x}{3} + \dfrac{4x}{5} = 11$

b $\dfrac{x+1}{3} + \dfrac{x+3}{2} = 10$

c $\dfrac{2x-5}{2} - \dfrac{x-1}{3} = 1$

4 Simplify each of these.

a $\dfrac{3x}{2} \times \dfrac{4x}{5}$

b $\dfrac{x+1}{4} \times \dfrac{3}{2x+2}$

c $\dfrac{2x-1}{2} \times \dfrac{4}{3x-1}$

5 Simplify each of these.

a $\dfrac{x}{4} \div \dfrac{2x}{5}$

b $\dfrac{x+3}{2} \div \dfrac{2x+6}{5}$

c $\dfrac{4x-2}{3} \div \dfrac{2x-1}{4}$

6 Show that $\dfrac{3}{x+2} + \dfrac{5}{2x-1} = 2$ simplifies to $4x^2 - 5x - 11 = 0$.

7 Solve these equations.

a $\dfrac{3}{x-1} + \dfrac{2}{2x+3} = 5$

b $\dfrac{5}{3x+2} - \dfrac{3}{2x-3} = 4$

c $\dfrac{5}{x+3} + \dfrac{2}{2x+6} = 4$

8 Simplify this expression. $\dfrac{x^2 - 2x - 3}{2x^2 - 10x + 12}$

24.2 Changing the subject of a formula

Homework 24B

1 Make the letter in brackets the subject of the formula.

a $4(x - 2y) = 3(2x - y)$ (x)

b $p(a - b) = q(a + b)$ (a)

c $A = 2ab^2 + ac$ (a)

d $s(t + 1) = 2r + 3$ (r)

e $st - r = 2r - 3t$ (t)

2 Make x the subject of each formula.

a $ax = b - cx$

b $x(a - b) = x + b$

c $a - bx = dx - a$

d $x(c - d) = c(d - x)$

3 **a** The perimeter of the shape on the right is given by the formula
$P = 2\pi r + 4r$.

Make r the subject of the formula.

b The area of the same shape is given by $A = \pi r^2 + 4r^2$.

Make r the subject of this formula.

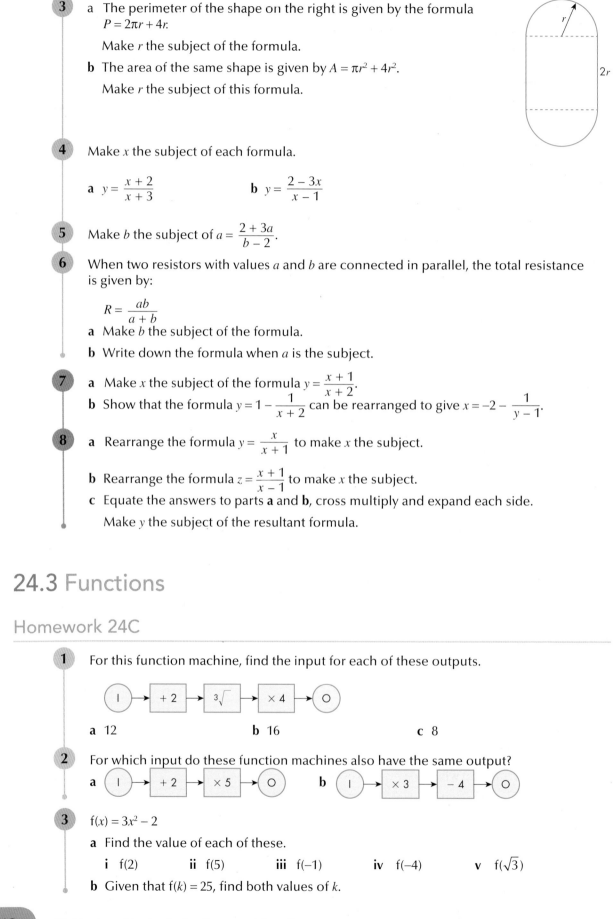

4 Make x the subject of each formula.

a $y = \dfrac{x + 2}{x + 3}$ **b** $y = \dfrac{2 - 3x}{x - 1}$

5 Make b the subject of $a = \dfrac{2 + 3a}{b - 2}$.

6 When two resistors with values a and b are connected in parallel, the total resistance is given by:

$R = \dfrac{ab}{a + b}$

a Make b the subject of the formula.

b Write down the formula when a is the subject.

7 **a** Make x the subject of the formula $y = \dfrac{x + 1}{x + 2}$.

b Show that the formula $y = 1 - \dfrac{1}{x + 2}$ can be rearranged to give $x = -2 - \dfrac{1}{y - 1}$.

8 **a** Rearrange the formula $y = \dfrac{x}{x + 1}$ to make x the subject.

b Rearrange the formula $z = \dfrac{x + 1}{x - 1}$ to make x the subject.

c Equate the answers to parts **a** and **b**, cross multiply and expand each side.

Make y the subject of the resultant formula.

24.3 Functions

Homework 24C

1 For this function machine, find the input for each of these outputs.

$$ \boxed{I} \rightarrow \boxed{+2} \rightarrow \boxed{\sqrt[3]{}} \rightarrow \boxed{\times 4} \rightarrow \boxed{O} $$

a 12   **b** 16   **c** 8

2 For which input do these function machines also have the same output?

a $\boxed{I} \rightarrow \boxed{+2} \rightarrow \boxed{\times 5} \rightarrow \boxed{O}$ **b** $\boxed{I} \rightarrow \boxed{\times 3} \rightarrow \boxed{-4} \rightarrow \boxed{O}$

3 $f(x) = 3x^2 - 2$

a Find the value of each of these.

   i $f(2)$ **ii** $f(5)$ **iii** $f(-1)$ **iv** $f(-4)$ **v** $f(\sqrt{3})$

b Given that $f(k) = 25$, find both values of k.

4 $g(x) = 8 - x^2$

    a Find the value of each of these.

        i $g(2)$ **ii** $g(-3)$ **iii** $g(6)$ **iv** $g(-4)$ **v** $g(\sqrt{7})$ **vi** $g(-0.5)$

    b Solve $g(x) = -1$.

5 $f(x) = 2x^2 - 6x + 4$

    a Find the value of each of these.

        i $f(-1)$ **ii** $f(5)$ **iii** $f(-2)$

    b Solve $f(x) = 0$.

Homework 24D

1 Find an expression for $f^{-1}(x)$ for each function.

    a $f(x) = 10x - 1$ **b** $f(x) = \frac{1}{3}x + 4$ **c** $f(x) = -5x - 10$

    d $f(x) = (x - 3)^2$ **e** $f(x) = \sqrt{x - 4}$

2 Given that $f(x) = \dfrac{x + 3}{2x - 1}$, find an expression for $f^{-1}(x)$.

3 Given that $f(x) = \dfrac{3x + 5}{2x - 3}$, find an expression for $f^{-1}(x)$.

4 Given that $f(x) = \dfrac{4x + 3}{x - 4}$, find an expression for $f^{-1}(x)$.

5 What do you notice about your answers to questions **2**, **3** and **4**? Make up a question of this type and find the inverse.

24.4 Composite functions

Homework 24E

1 Evaluate each composite value.

    a Find fg(3) when $f(x) = 3x - 5$ and $g(x) = x^2$.

    b Find fg(10) when $f(x) = -9x - 9$ and $g(x) = \sqrt{(x - 9)}$.

    c Find fg(12) when $f(x) = -4x + 2$ and $g(x) = \sqrt{(x - 8)}$.

    d Find gf(-2) when $f(x) = -3x + 4$ and $g(x) = x^2$.

    e Find gf(2) when $f(x) = -2x + 1$ and $g(x) = \sqrt{x^2 - 5}$.

2 Find each composite function.

    a Find fg(x) when $f(x) = -9x + 3$ and $g(x) = x^4$.

    b Find fg(x) when $f(x) = 2x - 5$ and $g(x) = x + 2$.

    c Find fg(x) when $f(x) = x^2 + 7$ and $g(x) = x - 3$.

    d Find gf(x) when $f(x) = 4x + 3$ and $g(x) = x^2$.

    e Find gf(x) when $f(x) = x - 1$ and $g(x) = x^2 + 2x - 8$.

3 A teacher asks his class to find the value of $fg(-3)$ when $f(x) = x^2 - 3$ and $g(x) = 5x$.
This is Wayne's answer.

$f(-3) = (-3)^2 - 3$	$g(x) = 5x$	$fg(-3) = 6(-15)$
$f(-3) = 9 - 3$	$g(-3) = 5(-3)$	$= -90$
$f(-3) = 6$	$g(-3) = -15$	

What has he done wrong? Solve the problem correctly.

24.5 Iteration

Homework 24F 🖩

1 Find the first five iterations of each iterative formulae. Start each one with $x_1 = 3$.

    a $x_{n+1} = \dfrac{x_n + 2}{6}$ **b** $x_{n+1} = \dfrac{x_n}{5} + 4$ **c** $x_{n+1} = \dfrac{2}{x_n - 5}$

2 Find a root of the quadratic equation $2x^2 + 3x - 9 = 0$ using the iterative formula:

$$x_{n+1} = \sqrt{\dfrac{9 - 3x_n}{2}}.$$

Start with $x_1 = 2$ and find a solution correct to 2 decimal places.

3 **a** Show that $x^2 + x - 1 = 0$ can be rearranged into the iterative formula $x_{n+1} = \sqrt{1 - x_n}$.

    b Use the iterative formula and a starting value of $x_1 = 0.5$ to obtain a root of the equation correct to 2 decimal places.

4 **a** Show that $x^2 - 9x + 2 = 0$ can be rearranged into the iterative formula
$x_{n+1} = \sqrt{9x_n - 2}$.

    b Use the iterative formula and a starting value of $x_n = 8$ to obtain a root of the equation correct to 2 decimal places.

5 A rectangle has sides of $(x - 3)$ cm and $(x + 4)$ cm and an area of 26 cm^2.

    a Show that $x^2 + x - 38 = 0$.

    b Use the iterative, formula $x_{n+1} = \sqrt{38 - x_n}$ and an initial input of $x_1 = 3$ to find the length of each side of the rectangle, correct to 2 decimal places.

6 **a** Show that $x = \dfrac{5}{x} - 3$ can be rearranged into the equation $x^2 + 3x - 5 = 0$.

    b Use the iterative formula $x_{n+1} = \dfrac{5}{x_n} - 3$ to find a root of the equation giving your answer to 2 decimal places.

7 Solve the equation $x^3 - 2x + 3 = 5$ using an iterative formula.

25 Geometry and measures: Vector geometry

25.1 Properties of vectors

Homework 25A

1 On this grid, $\vec{OA}$ is **a** and $\vec{OB}$ is **b**.

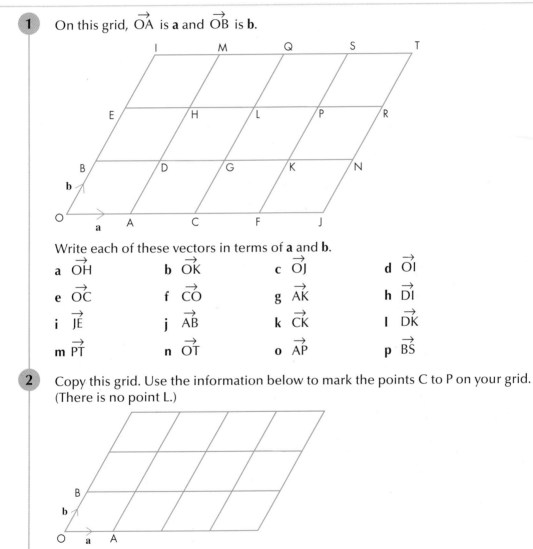

Write each of these vectors in terms of **a** and **b**.

a $\vec{OH}$	**b** $\vec{OK}$	**c** $\vec{OJ}$	**d** $\vec{OI}$
e $\vec{OC}$	**f** $\vec{CO}$	**g** $\vec{AK}$	**h** $\vec{DI}$
i $\vec{JE}$	**j** $\vec{AB}$	**k** $\vec{CK}$	**l** $\vec{DK}$
m $\vec{PT}$	**n** $\vec{OT}$	**o** $\vec{AP}$	**p** $\vec{BS}$

2 Copy this grid. Use the information below to mark the points C to P on your grid. (There is no point L.)

a $\vec{OC} = \mathbf{a} + \mathbf{b}$ **b** $\vec{OD} = 2\mathbf{a} + \mathbf{b}$ **c** $\vec{OE} = 2\mathbf{a}$ **d** $\vec{OF} = 3\mathbf{a}$

e $\vec{OG} = 3\mathbf{a} + \mathbf{b}$ **f** $\vec{OH} = 4\mathbf{a}$ **g** $\vec{OI} = 4\mathbf{a} + \mathbf{b}$ **h** $\vec{OJ} = 2\mathbf{b}$

i $\vec{OK} = \mathbf{a} + 2\mathbf{b}$ **j** $\vec{OM} = 3\mathbf{a} + 2\mathbf{b}$ **k** $\vec{ON} = \frac{3}{2}\mathbf{a} + 2\mathbf{b}$ **l** $\vec{OP} = \frac{1}{2}\mathbf{a} + \frac{3}{2}\mathbf{b}$

3 $\vec{OA}$ = **a** and $\vec{OB}$ = **b**. M is the midpoint of AB.

a Express these vectors in terms of **a** and **b**.

   i $\vec{AB}$ **ii** $\vec{AM}$ **iii** $\vec{OM}$

b i Draw on a copy of the diagram, the points X and Y such that $\vec{OX}$ = 2**a** + **b** and $\vec{OY}$ = **a** + 2**b**.

   ii Express $\vec{XY}$ in terms of **a** and **b**.

c What other vector on the diagram is equivalent to $\vec{XY}$?

4 OACB is a trapezium. $\vec{OA}$ = **a**, $\vec{OB}$ = **b** and $\vec{BC}$ = 2**a**. P and Q are the midpoints of $\vec{OB}$ and $\vec{AC}$.

a Express these vectors in terms of **a** and **b**.

   i $\vec{OP}$ **ii** $\vec{AQ}$ **iii** $\vec{PQ}$

b How can you tell that $\vec{PQ}$ is parallel to $\vec{OA}$?

5 $\vec{OA}$ = **a** and $\vec{OB}$ = **b**. Point C divides the line AB in the ratio 3 : 1.

a Express $\vec{OC}$ in terms of **a** and **b**.

b If point D is the midpoint of AC, express $\vec{OD}$ in terms of **a** and **b**.

6 $\vec{OA}$ = 10**q** and $\vec{OB}$ = 5**p**.

$\vec{AX}$ = 4$\vec{XB}$.

Write these vectors in terms of **p** and **q**.

a $\vec{AB}$ **b** $\vec{AX}$ **c** $\vec{OX}$

Not to scale

7 ABCDEF is a regular hexagon with centre O.

$\vec{OA}$ = **a** and $\vec{OB}$ = **b**.

a Write each of these vectors in terms of **a** and **b**. Give your answers in their simplest form.

   i $\vec{AB}$ **ii** $\vec{AD}$ **iii** $\vec{EC}$ **iv** $\vec{FB}$

b Write down two facts about the lines EC and FB.

8 A, B and C are three points. $\overrightarrow{AB} = 6\mathbf{a} + 4\mathbf{b}$ and $\overrightarrow{AC} = 9\mathbf{a} + 6\mathbf{b}$.

   a Write down one fact about the points A, B and C. Give a reason to support your answer.

   b Write down the ratio of the lengths AB : BC in its simplest form.

25.2 Vectors in geometry

Homework 25B

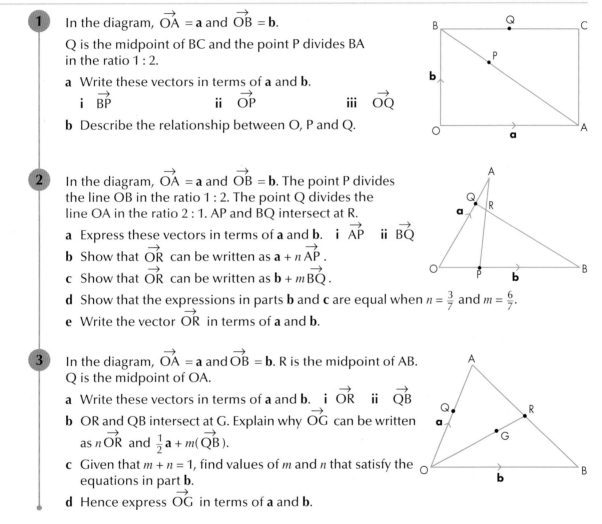

1 In the diagram, $\overrightarrow{OA} = \mathbf{a}$ and $\overrightarrow{OB} = \mathbf{b}$.

Q is the midpoint of BC and the point P divides BA in the ratio 1 : 2.

   a Write these vectors in terms of $\mathbf{a}$ and $\mathbf{b}$.

     i $\overrightarrow{BP}$ **ii** $\overrightarrow{OP}$ **iii** $\overrightarrow{OQ}$

   b Describe the relationship between O, P and Q.

2 In the diagram, $\overrightarrow{OA} = \mathbf{a}$ and $\overrightarrow{OB} = \mathbf{b}$. The point P divides the line OB in the ratio 1 : 2. The point Q divides the line OA in the ratio 2 : 1. AP and BQ intersect at R.

   a Express these vectors in terms of $\mathbf{a}$ and $\mathbf{b}$.   **i** $\overrightarrow{AP}$ **ii** $\overrightarrow{BQ}$

   b Show that $\overrightarrow{OR}$ can be written as $\mathbf{a} + n\overrightarrow{AP}$.

   c Show that $\overrightarrow{OR}$ can be written as $\mathbf{b} + m\overrightarrow{BQ}$.

   d Show that the expressions in parts **b** and **c** are equal when $n = \frac{3}{7}$ and $m = \frac{6}{7}$.

   e Write the vector $\overrightarrow{OR}$ in terms of $\mathbf{a}$ and $\mathbf{b}$.

3 In the diagram, $\overrightarrow{OA} = \mathbf{a}$ and $\overrightarrow{OB} = \mathbf{b}$. R is the midpoint of AB. Q is the midpoint of OA.

   a Write these vectors in terms of $\mathbf{a}$ and $\mathbf{b}$.   **i** $\overrightarrow{OR}$ **ii** $\overrightarrow{QB}$

   b OR and QB intersect at G. Explain why $\overrightarrow{OG}$ can be written as $n\overrightarrow{OR}$ and $\frac{1}{2}\mathbf{a} + m(\overrightarrow{QB})$.

   c Given that $m + n = 1$, find values of m and n that satisfy the equations in part **b**.

   d Hence express $\overrightarrow{OG}$ in terms of $\mathbf{a}$ and $\mathbf{b}$.

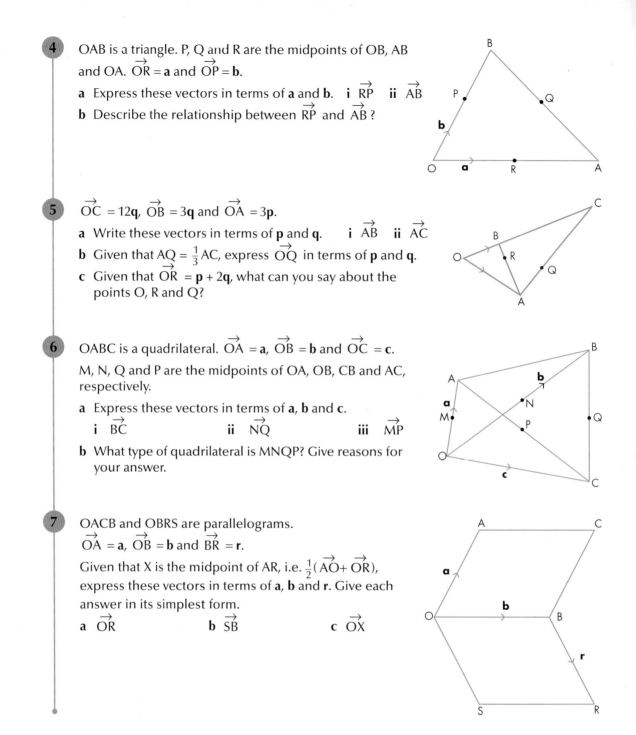

4 OAB is a triangle. P, Q and R are the midpoints of OB, AB and OA. $\overrightarrow{OR} = \mathbf{a}$ and $\overrightarrow{OP} = \mathbf{b}$.

   a Express these vectors in terms of **a** and **b**.   **i** $\overrightarrow{RP}$ **ii** $\overrightarrow{AB}$

   b Describe the relationship between $\overrightarrow{RP}$ and $\overrightarrow{AB}$?

5 $\overrightarrow{OC} = 12\mathbf{q}$, $\overrightarrow{OB} = 3\mathbf{q}$ and $\overrightarrow{OA} = 3\mathbf{p}$.

   a Write these vectors in terms of **p** and **q**.   **i** $\overrightarrow{AB}$ **ii** $\overrightarrow{AC}$

   b Given that $AQ = \frac{1}{3}AC$, express $\overrightarrow{OQ}$ in terms of **p** and **q**.

   c Given that $\overrightarrow{OR} = \mathbf{p} + 2\mathbf{q}$, what can you say about the points O, R and Q?

6 OABC is a quadrilateral. $\overrightarrow{OA} = \mathbf{a}$, $\overrightarrow{OB} = \mathbf{b}$ and $\overrightarrow{OC} = \mathbf{c}$.

   M, N, Q and P are the midpoints of OA, OB, CB and AC, respectively.

   a Express these vectors in terms of **a**, **b** and **c**.

     i $\overrightarrow{BC}$ **ii** $\overrightarrow{NQ}$ **iii** $\overrightarrow{MP}$

   b What type of quadrilateral is MNQP? Give reasons for your answer.

7 OACB and OBRS are parallelograms.

   $\overrightarrow{OA} = \mathbf{a}$, $\overrightarrow{OB} = \mathbf{b}$ and $\overrightarrow{BR} = \mathbf{r}$.

   Given that X is the midpoint of AR, i.e. $\frac{1}{2}(\overrightarrow{AO} + \overrightarrow{OR})$, express these vectors in terms of **a**, **b** and **r**. Give each answer in its simplest form.

   a $\overrightarrow{OR}$ **b** $\overrightarrow{SB}$ **c** $\overrightarrow{OX}$

8 In the diagram, $\overrightarrow{OA} = \mathbf{a}$, $\overrightarrow{OB} = \mathbf{b}$ and $\overrightarrow{OC} = 3\mathbf{b} - 2\mathbf{a}$.

Prove that ABC is a straight line.

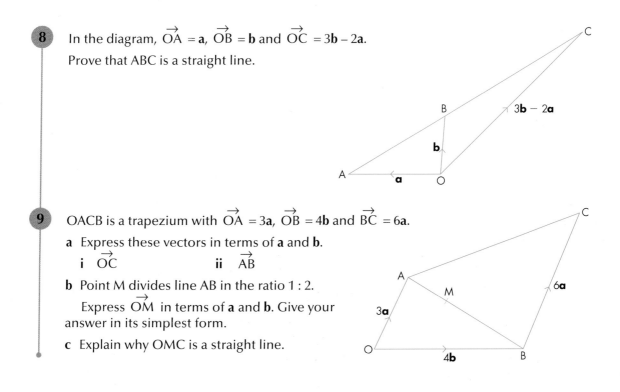

9 OACB is a trapezium with $\overrightarrow{OA} = 3\mathbf{a}$, $\overrightarrow{OB} = 4\mathbf{b}$ and $\overrightarrow{BC} = 6\mathbf{a}$.

   a Express these vectors in terms of **a** and **b**.

      i $\overrightarrow{OC}$ **ii** $\overrightarrow{AB}$

   b Point M divides line AB in the ratio 1 : 2.

      Express $\overrightarrow{OM}$ in terms of **a** and **b**. Give your answer in its simplest form.

   c Explain why OMC is a straight line.

William Collins' dream of knowledge for all began with the publication of his first book in 1819. A self-educated mill worker, he not only enriched millions of lives, but also founded a flourishing publishing house. Today, staying true to this spirit, Collins books are packed with inspiration, innovation and practical expertise. They place you at the centre of a world of possibility and give you exactly what you need to explore it.

Collins. Freedom to teach

Published by Collins
An imprint of HarperCollins*Publishers*
1 London Bridge Street
London SE1 9GF

Browse the complete Collins catalogue at
www.collins.co.uk

© HarperCollins*Publishers* Limited 2015

10 9 8 7 6 5 4 3 2 1

ISBN 978-0-00-811383-4

A catalogue record for this book is available from the British Library

The author Rob Ellis asserts his moral rights to be identified as the author of this work.

Commissioned by Lucy Rowland and Katie Sergeant
Project managed by Elektra Media Ltd and Hart McLeod Ltd
Copyedited by Marie Taylor
Proofread and answers checked by Amanda Dickson and Steven Matchett
Edited by Jennifer Yong
Typeset by Jouve India Private Limited
Illustrations by Ann Paganuzzi
Designed by Ken Vail Graphic Design
Cover design by We are Laura
Production by Rachel Weaver

Printed in Italy by Grafica Veneta S.p.A.

Acknowledgements
The publishers gratefully acknowledge the permissions granted to reproduce copyright material in this book. Every effort has been made to contact the holders of copyright material, but if any have been inadvertently overlooked, the publisher will be pleased to make the necessary arrangements at the first opportunity.

The publishers would like to thank the following for permission to reproduce photographs in these pages:

Cover (bottom) Procy/Shutterstock, cover (top) joingate/Shutterstock, p 121 digidreamgrafix/Shutterstock.